Ultrasonic Cavitation Treatment of Metallic Alloys

Ultrasonic Cavitation Treatment of Metallic Alloys

Special Issue Editors

Iakovos Tzanakis
Dmitry Eskin

MDPI • Basel • Beijing • Wuhan • Barcelona • Belgrade

Special Issue Editors
Iakovos Tzanakis
Oxford Brookes University
UK

Dmitry Eskin
Brunel University London
UK

Editorial Office
MDPI
St. Alban-Anlage 66
4052 Basel, Switzerland

This is a reprint of articles from the Special Issue published online in the open access journal *Materials* (ISSN 1996-1944) from 2019 to 2020 (available at: https://www.mdpi.com/journal/materials/special_issues/ultrasonic_metallic_alloys)

For citation purposes, cite each article independently as indicated on the article page online and as indicated below:

LastName, A.A.; LastName, B.B.; LastName, C.C. Article Title. *Journal Name* **Year**, *Article Number*, Page Range.

ISBN 978-3-03928-196-1 (Pbk)
ISBN 978-3-03928-197-8 (PDF)

© 2020 by the authors. Articles in this book are Open Access and distributed under the Creative Commons Attribution (CC BY) license, which allows users to download, copy and build upon published articles, as long as the author and publisher are properly credited, which ensures maximum dissemination and a wider impact of our publications.

The book as a whole is distributed by MDPI under the terms and conditions of the Creative Commons license CC BY-NC-ND.

Contents

About the Special Issue Editors vii

Preface to "Ultrasonic Cavitation Treatment of Metallic Alloys" ix

Nagasivamuni Balasubramani, David StJohn, Matthew Dargusch and Gui Wang
Ultrasonic Processing for Structure Refinement: An Overview of Mechanisms and Application of the Interdependence Theory
Reprinted from: *Materials* **2019**, *12*, 3187, doi:10.3390/ma12193187 1

Koulis Pericleous, Valdis Bojarevics, Georgi Djambazov, Agnieszka Dybalska, William D. Griffiths and Catherine Tonry
Contactless Ultrasonic Cavitation in Alloy Melts
Reprinted from: *Materials* **2019**, *12*, 3610, doi:10.3390/ma12213610 32

Tao Zeng and YaJun Zhou
Effects of Ultrasonic Introduced by L-Shaped Ceramic Sonotrodes on Microstructure and Macro-Segregation of 15t AA2219 Aluminum Alloy Ingot
Reprinted from: *Materials* **2019**, *12*, 3162, doi:10.3390/ma12193162 45

Sergey Komarov and Takuya Yamamoto
Role of Acoustic Streaming in Formation of Unsteady Flow in Billet Sump during Ultrasonic DC Casting of Aluminum Alloys
Reprinted from: *Materials* **2019**, *12*, 3532, doi:10.3390/ma12213532 58

Bruno Lebon, Iakovos Tzanakis, Koulis Pericleous and Dmitry Eskin
Numerical Modelling of the Ultrasonic Treatment of Aluminium Melts: An Overview of Recent Advances
Reprinted from: *Materials* **2019**, *12*, 3262, doi:10.3390/ma12193262 74

Chuangnan Wang, Thomas Connolley, Iakovos Tzanakis, Dmitry Eskin and Jiawei Mi
Characterization of Ultrasonic Bubble Clouds in A Liquid Metal by Synchrotron X-ray High Speed Imaging and Statistical Analysis
Reprinted from: *Materials* **2020**, *13*, 44, doi:10.3390/ma13010044 87

About the Special Issue Editors

Iakovos Tzanakis (Dr.) is a Reader in Engineering Materials at Oxford Brookes University, Oxford, UK. Dr Tzanakis received his Ph.D. from Bournemouth University (UK) in 2010 and then worked as a Research Fellow in Bournemouth (2010–2012) and Brunel University London (2013–2016). From April 2016, Dr. Tzanakis was appointed as a Lecturer in Engineering Materials at Oxford Brookes University, gaining promotions to Senior Lecturer (2017) and Reader in 2018. In 2017, he was granted access via a H2020 call to the advanced NFFA-Europe facilities for the sono-exfoliation of 2D functional materials. He is currently a Principal Investigator for the EPSRC projects UltraMelt2 (EP/R011044/1) and EcoUltra2D (EP/R031401/1) with a total budget of £700k. Dr. Tzanakis has unique expertise in the fundamental research of cavitation and advanced characterization of ultrasonic liquid processing and cavitation erosion, including the application of an advanced cavitometer for acoustic measurements and PIV to the study of acoustic flows as well as in situ ultra-high-speed imaging and synchrotron studies. He has over 45 papers in high-ranking journals to his name (h-index 16).

Dmitry G. Eskin (Prof.) is Professor at Brunel Centre for Advanced Solidification Technology (BCAST) at Brunel University London, UK. Prof. Eskin received his Ph.D. degree from Moscow Institute of Steel and Alloys in 1988, and worked at senior academic positions in the Netherlands (1999–2010, TU Delft, M2i, NIMR) and Russia (1988–1999, Inst. of Metallurgy). He has published >250 papers and authored/co-authored 7 monographs including Multicomponent Phase Diagrams: Applications for Aluminum Alloys (2005), Physical Metallurgy of Direct-Chill Casting (2008), and Ultrasonic Treatment of Light Alloy Melts (2015) (>7300 citations, h-index 41 with books) concerning solidification processing, alloy development, and ultrasonic processing. In the last 7 years, he supervised 5 Ph.D. students and was Principal Investigator of 8 projects, including 3 FP7 projects on ultrasonic processing of composite materials and liquid metals and four EPSRC projects—UltraMelt (EP/K005804/1), UltraCast (EP/L019884/1), UltraMelt2 (EP/R011095/1), EcoUltra2D (EP/R031665/1)—studying the interaction of ultrasonic cavitation with liquid metal flow and solid interfaces, with a total research budget of £2.8M. Prof. Eskin is a leading specialist in the fundamentals of solidification and ultrasonic processing with major contributions to modern views on the mechanisms of casting defect formation and on ultrasonic melt processing. Prof. Eskin is an editor of *J. Alloys and Compounds*, subject editor of JOM, and has received the TMS Warren Peterson Cast Shop for Aluminum Production Awards (2011, 2013), TMS Aluminum Technology Award (2013), and Mendeleev Medal from TSU (2018).

Preface to "Ultrasonic Cavitation Treatment of Metallic Alloys"

The current trend in solidification research is to develop a generic, energy-efficient, economical, sustainable, and pollution-free technology that can be applied to different alloy systems. Ultrasonic-cavitation melt treatment (UST) is a rather universal technology that can be applied to conventional and advanced metallic materials, regardless of their composition, while being environmentally friendly, cost effective, and ready to be implemented in conventional casting technologies such as direct-chill, continuous, or shape casting, as well as in emerging technologies of additive manufacturing and nanocomposite materials.

The beneficial effects of UST—such as in assisted nucleation, activation of substrates (wetting), deagglomeration and fragmentation of solid phases, degassing of the melt, and grain refinement of the as-cast product—stem from the growth, collapse, and implosion of cavitation bubbles as a result of alternate fluctuations in ultrasonic pressure. Although successfully demonstrated on the laboratory and pilot scale, UST has not yet found widespread industrial implementation. This is mostly due to the lack of in-depth understanding of the fundamental mechanisms behind the improved metal quality and structure refinement.

Thus, fundamental research is needed to answer the following practical questions: What is the optimum melt flow rate that maximizes treatment efficiency whilst minimizing input power, cost, and plant complexity? What is the optimum operating frequency and acoustic power that accelerates the treatment effects? What is the optimum location of an ultrasonic power source in the melt transfer system in relation to the melt pool geometry? Answering these questions will pave the way for widespread industrial use of ultrasonic melt processing with the benefit of improving the properties of lightweight structural alloys, simultaneously alleviating the present use of polluting (Cl, F) for degassing or expensive (Zr, Ti, B, Ar) grain refinement additives.

Iakovos Tzanakis, Dmitry Eskin
Special Issue Editors

Article

Ultrasonic Processing for Structure Refinement: An Overview of Mechanisms and Application of the Interdependence Theory

Nagasivamuni Balasubramani [1], David StJohn [1,2], Matthew Dargusch [1,2] and Gui Wang [1,2,*]

[1] Centre for Advanced Material Processing and Manufacturing (AMPAM), School of Mechanical and Mining Engineering, The University of Queensland, St Lucia 4072, Australia; n.balasubramani@uq.edu.au (N.B.); d.stjohn@uq.edu.au (D.S.); m.dargusch@uq.edu.au (M.D.)
[2] DMTC Limited, The University of Queensland, St Lucia 4072, Australia
* Correspondence: gui.wang@uq.edu.au

Received: 6 September 2019; Accepted: 26 September 2019; Published: 28 September 2019

Abstract: Research on ultrasonic treatment (UST) of aluminium, magnesium and zinc undertaken by the authors and their collaborators was stimulated by renewed interest internationally in this technology and the establishment of the ExoMet program of which The University of Queensland (UQ) was a partner. The direction for our research was driven by a desire to understand the UST parameters that need to be controlled to achieve a fine equiaxed grain structure throughout a casting. Previous work highlighted that increasing the growth restriction factor Q can lead to significant refinement when UST is applied. We extended this approach to using the Interdependence model as a framework for identifying some of the factors (e.g., solute and temperature gradient) that could be optimised in order to achieve the best refinement from UST for a range of alloy compositions. This work confirmed established knowledge on the benefits of both liquid-only treatment and the additional refinement when UST is applied during the nucleation stage of solidification. The importance of acoustic streaming, treatment time and settling of grains were revealed as critical factors in achieving a fully equiaxed structure. The Interdependence model also explained the limit to refinement obtained when nanoparticle composites are treated. This overview presents the key results and mechanisms arising from our research and considers directions for future research.

Keywords: ultrasonic treatment; grain refinement; interdependence model; aluminium alloys; magnesium alloys; zinc

1. Introduction

Our interest in the potential of UST to refine the grain structure of metals began in 2008 when it was realised that alloy composition can have a significant effect on grain size even under the application of UST [1]. It was found that grain size is linearly related to the inverse of the growth restriction factor Q in line with observations for alloys cast by traditional casting processes [1–3]. The other incentive was the increase in interest shown by industry and academia to use UST to potentially refine the grain size to below that achieved by the addition of inoculant particles, or to eliminate the use of inoculant containing master alloys such as Mg-Zr for Al-free magnesium alloys which are expensive and wasteful [4–6]. Further stimulus came from partnering with the European project ExoMet in 2012 that focused on the development of new liquid metal processing techniques using external fields (mainly UST) on Mg and Al based light alloys and nanocomposites. The initial focus of our research was to understand the role of alloy composition in relation to the effectiveness of UST as a grain refinement method. In the process we gained knowledge about the mechanisms and parameters that need to be controlled to ensure UST provides consistent grain refinement throughout a casting. Thus, after a decade of research on a number of alloy systems and compositions, it is a good time to review this work as a whole with some new experimental results in the expectation that further insights will be revealed.

Controlling the as-cast grain size of castings is one of the important steps for achieving the desired quality and mechanical properties of a cast product [7,8]. Fine, equiaxed and non-dendritic grain structures in castings provide uniform mechanical properties, improved distribution of secondary phases and enhanced feeding to eliminate shrinkage porosity [8–11]. In the case of ingot or continuous casting, a refined uniform grain structure can significantly improve the downstream processing ability and productivity such as forging, rolling and extrusion operations by reducing the risk of hot tearing and macro-segregation [9,11]. For instance, producing a fine grain microstructure in Mg alloys throughout the thickness of the sheet without composition variation is critical for manufacturing high quality sheet by twin-roll casting [12].

Generally, grain refinement is induced by the addition of a master alloy containing heterogeneous nucleating particles. For example, the addition of Ti and Zr to specific Al and Mg alloys produces excellent refinement in grain size of cast products [8,13,14]. Although the addition of grain refining master alloy is the most common foundry practice, there are several disadvantages including (i) low efficiency where only a few percent or even less of the added particles are active in grain nucleation [15,16]; (ii) refiners have only been identified for certain alloy systems [8,13,14]; (iii) difficulty in uniformly distributing the refiner particles into the melt without the formation of agglomerates; (iv) interaction of nucleant particles with impurity elements leading to a poisoning effect [17,18] or grain coarsening by the addition of the alloying element itself (e.g., Si poisoning in Al alloys) [19]; (v) fading due to the loss of the refinement ability by particle settling; and (v) the high cost of master alloys [5,6].

Liquid and semi-solid melt processing with power ultrasound has received considerable attention and has been successfully demonstrated for the refinement of as-cast grain structure in light alloys without the need for grain refiners, therefore eliminating the limitations associated with inoculation processes for Al and Mg based alloys [20–22]. The benefits of ultrasonic treatment (UST) include the production of non-dendritic structures, degasification and the refinement of primary phases (Al_3Zr, Al_3Ti, Al-Fe-Si and primary Si) [23–25]. This technique has attracted commercial interest due to being environmentally friendly, cost effective and possesses several technical advantages over conventional methods in molten metal processing [23,26,27].

Earlier work by Eskin [22,23] and Abramov [28] showed the potential of UST to refine the grain structure in various alloys and its possible mechanisms. The occurrence of cavitation (nucleation and implosion of bubbles during alternate pressure cycles of ultrasound) and acoustic streaming were identified as two key phenomena that change the dynamics of solidification. Significant effort has been made by Eskin and co-workers to understand the role of cavitation and acoustic steaming using advanced X-ray synchrotron solidification techniques and by modelling the dynamics of cavitation bubble formation [29–34]. A recent review by Eskin et al. [34] provided a broad summary of the effects of cavitation and acoustic streaming (mainly on in-situ observations and ex-situ solidification studies) on heterogeneous nucleation, fragmentation of primary crystals and intermetallic phases and de-agglomeration mechanisms. Alternative to the in-situ radiography techniques, observation of organic transparent analogues (e.g., succinonitrile (SCN)-1 wt.% camphor alloy) using a high-speed camera during UST also reveals the streaming flow pattern and the interaction of cavitation bubbles with dendrites and the mechanisms of fragmentation [35]. These in-situ solidification studies have increased our current understanding by elucidating the role of cavitation bubbles and explained the importance of these mechanisms for the upscaling of UST to large melt volumes in industrial applications. Considerable research has also highlighted the importance of cavitation and acoustic streaming on the activation of potent particles, nucleation of grains, altering convection patterns and reducing the temperature gradients during solidification [23,26–28,36–40]. While recent publications on advanced (in-situ) solidification studies have provided more detailed evidence on bubble dynamics during melt solidification [31–33,41,42], our research has been focused on other factors such as solute content, type of solute, constitutional supercooling, role of potent and impotent (oxide) particles, UST duration, origin and transport of grains, temperature range over which UST is applied, sonotrode preheating and other casting variables that could affect grain formation [6,37,38,43–47]. Using the Interdependence Theory of nucleation and grain refinement [15,48], we have revealed explanations for the role of solute [37,38,45,49], casting conditions, and micro [6,46] and nanoparticles [50] in assisting refinement by the application of UST.

This paper begins with a brief description of aspects of the experimental design such as the choice of metals, alloys and master alloys that were investigated, experimental techniques involved in the casting process, control of UST variables, modelling and simulation of acoustic streaming, and analytical methods. The results of these experiments are presented for two situations: UST applied above the liquidus temperature and UST applied from above to below the liquidus to include the nucleation stage of solidification. This is followed by an analysis of the origin and transport of grains, role of alloy elements and then an analysis of the grain nucleation mechanisms using the Interdependence model.

2. Experimental Design

2.1. Metals and Alloys Investigated

To investigate the effects of alloy composition, temperature range and duration of the application of UST, and the refinement mechanisms operating during solidification, different alloy systems were chosen from pure metals (Al, Mg, Zn), alloys (Al-Cu, Al-Si, Al-Mg) and alloys with the addition of master alloys: Al and its alloys with AlTiB master alloy, Mg and its alloys with Mg-Zr master alloy. The alloys and pure metal ingots were cast from commercial purity Al (99.7%), high purity Zn (99.995%), commercial purity Mg (99.91%), copper (99.9%) and silicon (99.4%). Al alloys with varying concentration of Si [45], Cu [37,51,52] and Mg [49] were cast with and without UST. Master alloys of Al3Ti1B and Mg-25Zr were introduced into the melt to understand the effect of nucleant particles [6,46]. Approximately 135 to 210 cc of pure metals and alloys were melted and solidified in a clay-graphite crucible in most of the experiments presented in [6,43,44,47,49,52]. The melt temperature was chosen to ensure there was enough superheat to overcome the chill effect induced by the unpreheated sonotrode. This ensured that the sonotrode was heated sufficiently to establish acoustic streaming in the melt before solidification began.

2.2. Application of UST

The experimental set-up is illustrated in Figure 1. An ultrasonic system (Sonics VCX1500, 20 kHz and 1.5 kW, Sonics & Materials, Inc., Newtown, CT, USA) made of a piezoelectric transducer, power generator, air cooling unit and an adjustment handle for lowering the sonotrode into the melt. A fixed power of 40% of the total power was used in all the UST experiments. The sonotrode was made of titanium alloy or molybdenum alloy designed for half a wavelength ($\lambda/2 = 125$ mm) having a diameter of 19 mm. After melting, the crucible containing the liquid metal from the furnace, was transferred to the platform and cooled in air as shown in Figure 1 [44]. The platform also has provision to measure the cooling curves using K-type thermocouples. The sonotrode (sometimes at room temperature or preheated to 285 °C) was turned on in air and then inserted into the melt at the required temperature, which then solidified in air for a specified duration. Table 1 shows the details of the pure metals and alloys investigated for refinement of grains and primary intermetallic phases classified according to the alloy type with a short summary of the outcomes of the work. Using the fixed value of ultrasonic power, grain refinement was studied with respect to temperature range and time duration of UST. For the casting conditions and alloys investigated in the present work, the UST duration was varied from 30 s to 4 min in Mg and Al castings and 2 to 9 min in Zn castings. To account for the effect of volume, three additional large volume castings (~322, 530, 946 cc) of Al-2Cu alloys were solidified under UST in order to understand the effect of casting volume and height (H < λ/2, H = λ/2 and H > λ/2) on the refinement of grain structure.

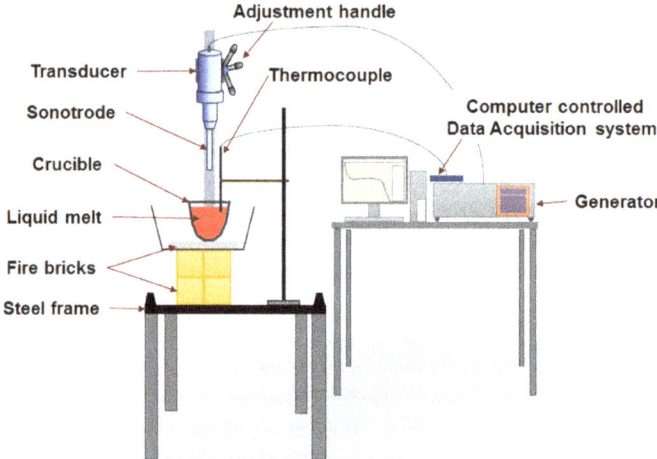

Figure 1. Ultrasonic treatment (UST) platform (steel frame) shows the arrangement of a crucible containing liquid melt placed over fire bricks. The adjustable handle is used to lower the sonotrode into the melt when the required temperature is recorded by the data acquisition system (K-type thermocouple).

Table 1. List of the experiments undertaken at The University of Queensland (UQ) and allied research groups at RMIT University and Helmholtz-Zentrum Geesthacht (HZG).

Alloy Type	Family of Pure Metals and Alloys	Composition (wt.%)	Summary of the Investigation Undertaken with UST	Ref.
Pure metals	Al	-	Evolution of grain structures with respect to temperature range and time duration by keeping the UST power constant	[46]
	Mg			[44]
	Zn			[43]
Eutectic alloys	Al-Cu	2, (1, 2, 3, 5, 7, 10)	Temperature range, time, solute effect and sonotrode preheating	[51,52]
	Al-Mg	5, 10, 15, 20	Role of solute, analysis of mechanism using IDM[a]	[49]
	Al-Si	1, 2.5, 4, 7, 10	Role of Si and its mechanism	[45]
Peritectic alloys	Al-Ti	0.005, 0.01, 0.02, 0.05, 0.1	Role of Ti, Si, Cu and TiB$_2$ particles, temperature range, mechanisms of refinement using IDM[a]	[46,47]
	Al-Si-Ti	Si (1, 2.5, 4, 7, 10), Ti (0.1)		[45]
	Al-Cu-Ti	Cu (2), Ti (0.005, 0.05, 0.1, 0.2)		[46,53]
	Al-Si-Fe[#]	Si (19), Fe (4)	Primary Si and Al-Fe-Si refinement, peritectic reactions	[24]
	Al-Si-Fe-Mn[#]	Si (17), Fe (2), Mn (0.5, 1, 1.5, 2)		[25]
	Mg-Zr	0.2, 0.4, 0.5, 0.8, 1.0	Settling tendency and size distribution of Zr particles, Zr dissolution and grain refinement mechanism using IDM[a]	[6]
Nanocomposites	AM60-1% AlN*	Al (6), Mn (0.4)	Grain refinement, mechanical properties, mechanism of refinement using IDM[a]	[50]

[#]RMIT University, Australia. *HZG, Germany. [a]Interdependence Model (IDM).

2.3. Macro and Micro Structure Examination

Solidified samples with and without UST were sectioned vertically to analyse the macrostructure and small samples of (15 mm × 10 mm) were taken to analyse the microstructural refinement. After polishing the samples using the standard metallographic preparation methods and etching, microstructures were examined in a polarised light microscope (Leica Polyvar). The grain size measurements were calculated using the linear intercept method by image analysis software [6,43]. The specific procedures of the composition analysis, sampling regions, etchants and etching techniques used for individual alloy systems can be found in the respective references listed in Table 1.

2.4. Modelling and Validation of Acoustic Streaming

Theoretically, acoustic streaming can be considered as a time averaged transfer of momentum flux per unit area that creates a non-uniform accelerating flow which converts the wave energy into fluid motion [54]. A jet of streaming flow during solidification could significantly affect solute diffusion, convection and transport phenomena at the solid-liquid interface, which in turn is associated with dendritic fragmentation, grain growth, instabilities and the morphological features of grains [55,56]. The flow pattern induced by the acoustic stream can be tracked visibly using transparent analogues such as water, glycerine and ethanol [57]. In the case of liquid melts, it is difficult to directly observe the flow pattern induced by acoustic streaming, however, with reference to the transparent analogues, numerical solvers can be used to predict acoustic streaming and its effect on temperature gradient, flow velocity and acoustic pressure gradients [37,38,58–62].

Acoustic streaming induced fluid flow during UST was modelled by Wang et al. [38] using momentum equations in the ProCAST simulator that drives the net acoustic forces generated by the radiating surface propagating into the liquid using the Reynolds turbulent wave equation. The model aimed to identify the impact of forced flow in the melt affecting the temperature distribution of Al-2Cu alloy solidified in clay graphite-crucible (210 cc volume). Specific dimensions and heat transfer coefficients can be found in the ref [38]. Acoustic streaming was calculated using Navier–Stokes model at high Reynolds number as

$$\rho(\vec{v} \cdot \nabla \vec{v}) = -\nabla \bar{p} + \mu \nabla^2 \vec{v} + \vec{F}_N \quad (1)$$

where, ρ is density, $\bar{p}$ is the mean pressure and μ is the kinematic viscosity. It is assumed that the powerful streaming was created beneath the sonotrode tip and forced into the liquid at a velocity v, attenuating at a constant rate (P) with increasing distance from the tip of the sonotrode. The net force acting on the fluid exerts a momentum force along a particular direction (x) and is given by

$$\vec{F}_N = \frac{P}{c}(1 - e^{-\beta x}) \quad (2)$$

Here β is the attenuation coefficient. The resultant kinematic momentum (K) is given by

$$K = \rho \cdot \vec{F}_N \quad (3)$$

A momentum source of approximately 2.3×10^{-6} m^3 located beneath the sonotrode induces a net force of 0.024 N along the principal direction of the acoustic streaming (vertically downwards). The convective motion is then modelled using Navier–Stokes equations coupled with energy equations [38]. Figure 2a shows the casting setup with two thermocouples placed at the mould wall (T/C1) and the off-centre position (T/C2) to validate the current model. The cooling curves in Figure 2b,c show a good correlation between the measured (T_m) and simulated (T_s) temperature measurements in the as-cast condition and after the application of UST. This also validates the current assumption regarding the attenuation rate (P/c) of the momentum source (K), which is used to simulate the temperature gradients in the casting during solidification. This modelling approach does not include the incorporation of the effect of cavitation, simulation of grain size as a result of fragmentation and grain transport due to acoustic streaming [37,38].

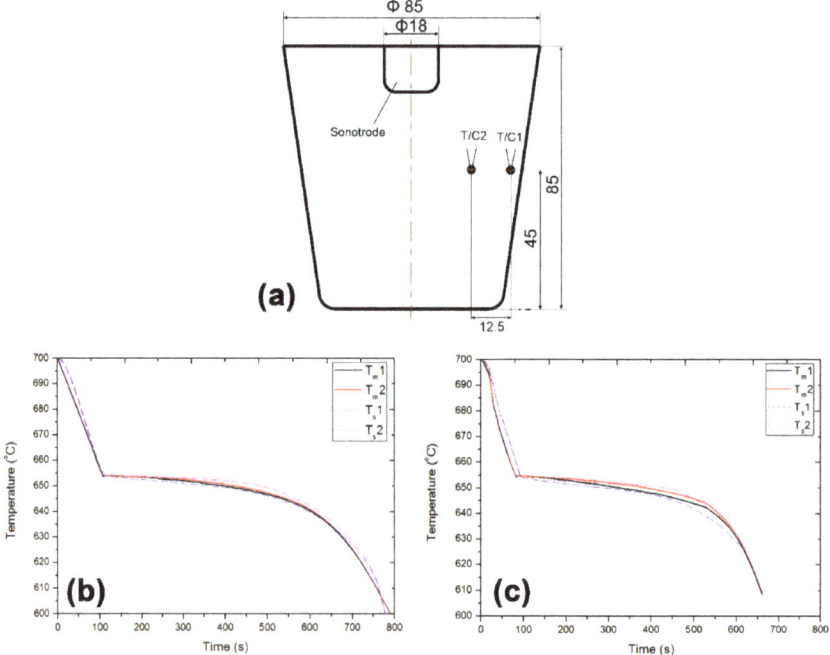

Figure 2. (a) Schematic of the casting setup with two thermocouples for the validation of the model and the corresponding cooling curves in (b) as-cast condition without UST and (c) after UST [38].

3. Simulation of Acoustic Streaming

Figure 3 shows the simulated velocity, solid fraction and temperature profile under a momentum induced acoustic streaming model. The simulation results showed that the fluid velocity varies from 0.38 m/s at the start of solidification producing a nearly flat temperature gradient and decreases to 0.32 m/s with the increase in solid fraction during solidification. The fluid velocity profile in Figure 3a shows a maximum velocity directly beneath the sonotrode and the recirculation pattern shows a high fluid velocity at the mould wall regions. Figure 3b shows the solid fraction profile corresponding to the velocity map shown in Figure 3a taken at the same solid fraction of 7.7%. It is revealed that a higher solid fraction contour is found at the centre region beneath the sonotrode, indicating that the first to form solid lies beneath the sonotode which gets pushed downstream by the action of acoustic streaming. Figure 3c shows the temperature distribution profile in as-cast and during UST conditions. Without UST, the mould walls of the crucible extract heat rapidly from the melt and the temperature distribution shows a steep gradient from the wall region to the centre of the crucible. After UST, the simulated profile shows almost a flat temperature with reduced gradient from the mould wall and the coldest zone being the cavitation zone where the grains are being produced and dispersed. By correlating the simulation results with the experimental grain size results, it was found that the high fluid velocity created by acoustic streaming tends to flatten the temperature gradient promoting the formation of an equiaxed grain structure [37,38].

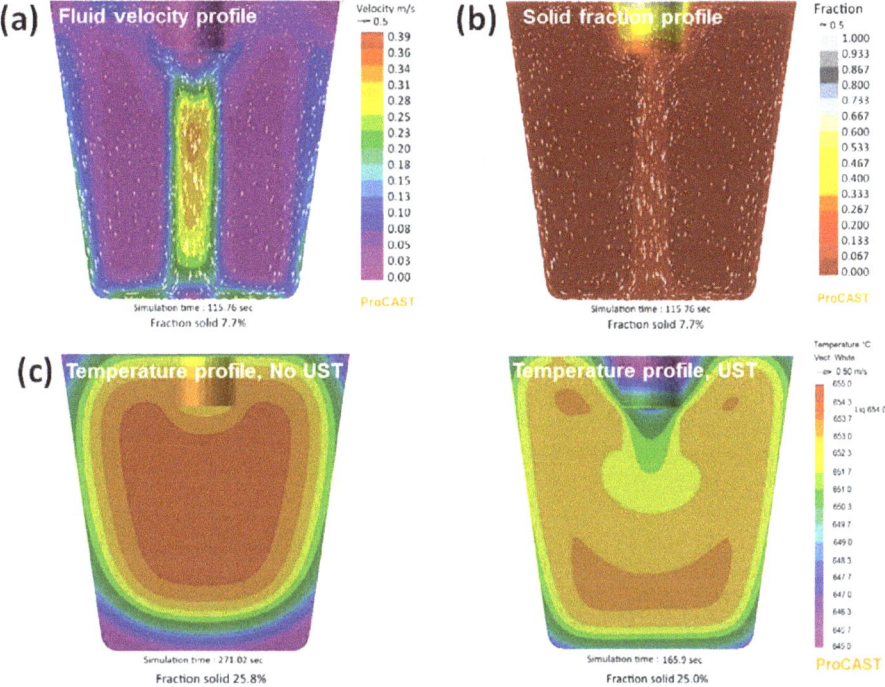

Figure 3. Simulations of (**a**) Fluid velocity and (**b**) solid fraction during UST, and (**c**) temperature gradient before and after UST in Al–2Cu alloy [38].

The flow fields affecting the temperature gradient can also be understood by cooling curve analysis during the solidification process. Figure 4a shows the simulated solidification profile at three points (A to C). The cooling curves from the corresponding points are shown in Figure 4b when the sonotrode is immersed into the melt but without turning on the sonotrode and Figure 4c where the sonotrode is turned on before immersion into the melt. After inserting the idle sonotrode into the melt the location at C and B is almost undisturbed while cooling curve A shows a significant drop in the temperature. Location A in Figure 4a represents the cavitation zone beneath the sonotrode where the observed temperature drop is 660 °C, which is only 5 °C above the liquidus temperature of the Al-2Cu alloy [52]. When the temperature drops below the liquidus the cavitation zone is referred to as undercooled zone, which is also confirmed by the simulation (Figure 3b,c). When UST is initiated, all the cooling profiles appear to be similar (Figure 4c). Cooling curve A (solid line) almost matches curve B after UST indicating that the bulk melt temperatures are the same allowing the powerful acoustic streaming to stabilise the temperature in A, B and C enabling the transport of the new grains formed beneath the sonotrode into the bulk melt (Figure 4c).

This stabilised bulk temperature reduces the risk of remelting grains and retains the non-dendritic structure after solidification.

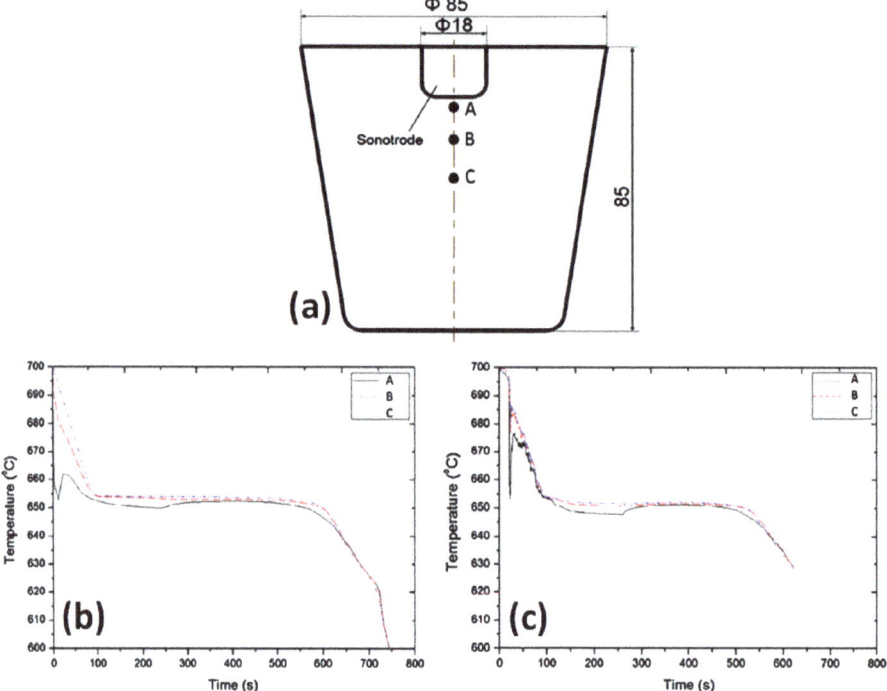

Figure 4. (a) Schematic of the casting setup shows the locations of points (A, B, C) and the simulated cooling curves of an Al-2Cu alloy when (b) the sonotrode is immersed into the melt but is not turned on during solidification and (c) the sonotrode is turned on before immersion into the melt and remains on during solidification as detailed in [38].

4. UST Applied to the Liquid Melt

In liquid melt, UST can be applied in two ways, either isothermally treated at a specific melt temperature and then poured into the mould [2,63,64] or treating the liquid melt over a temperature range during continuous cooling and terminated before the onset of primary phase nucleation at the liquidus temperature [6,47]. The melt above the liquidus temperature (superheated condition) always contains insoluble impurities (mainly oxides) that are not actively involved in nucleation under the normal casting conditions [23]. The occurrence of cavitation in the melt is advantageous by wetting these un-wetted particles turning them into active nucleation sites for grain refinement [23,27].

4.1. Pure Metals and Alloys

UST of pure Al does not result in significant refinement, however, it has been reported that Al-11Cu and Al-4Cu alloys showed approximately 20% to 25% refinement in the grain size. With the intentional addition of Al_2O_3 particles it was found that UST refinement in the liquid condition is governed by cavitation enhanced activation of heterogeneous oxide particles [2]. Figure 5 shows the macrostructures of pure Mg before and after UST terminated at 10 °C above and at the melting temperature respectively. Approximately 25% to 28% refinement is observed compared to the as-cast structure, however, the degree of refinement is not significant (Figure 5b,c). It is interesting to note that the grain orientation is non-uniform after UST, while in the as-cast condition the grains are oriented opposite to the direction of heat extraction. A similar observation has been reported for the Al-2Cu alloy when UST is applied from 714 °C to 660 °C (liquidus temperature = 655 °C) resulting in an insignificant reduction in grain size. The observed refinement could be associated with oxides or native impurity particles [44].

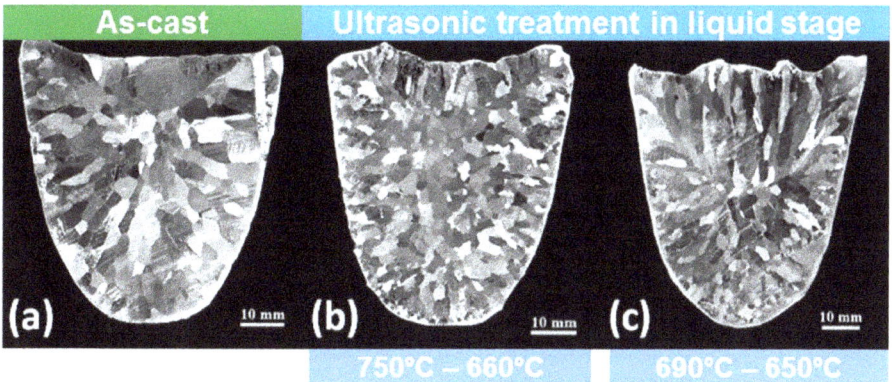

Figure 5. Macrostructure of commercial purity Mg in (**a**) without UST and after UST terminated at (**b**) 660 °C and (**c**) 650 °C [44].

4.2. With the Addition of Refining Master Alloys

Wang et al. [47] investigated the effect of adding Al$_3$Ti1B master alloy to pure Al by terminating UST at specific temperatures above the liquidus temperature during continuous cooling (from 720 °C to 660 °C). Figure 6a shows the grain refinement obtained after the addition of 50 and 200 ppm Ti with respect to the termination temperature of UST above the liquidus temperature. Commercial purity Al inoculated with 50 ppm Ti and 200 ppm in the as-cast condition reduces the grain size from the order of millimetres to ~300–150 µm respectively. Reducing the UST termination temperature close to the liquidus temperature further decreases the grain size at low refiner addition (50 ppm Ti). When the addition level is increased to 200 ppm, better refinement is achieved at higher temperatures. A significant difference in refinement is noted at 10 s UST at 700 °C, where 200 ppm of Ti produces excellent refinement compared to 50 ppm of Ti addition. This difference is reduced when the UST termination temperature reaches the liquidus temperature. For the known distribution of TiB$_2$ particles in the AlTiB master alloy and grain size relationship established with a range of wrought Al alloys [65], the dispersion of nucleant particles after the application of UST were characterised using the inter-particle distance (x_{sd}). It has been found that the grain size obtained after UST is much smaller than that predicted by x_{sd}, suggesting that a larger number of potent substrates are distributed homogeneously and simultaneously activated for better refinement [46,47].

Figure 6b shows the refinement of Mg-Zr alloys (plotted for nominal composition). Increasing the addition of Zr from 0.2 to 1.0 wt.% increases the liquidus temperature of the Mg-Zr alloys from 650 °C to 653 °C. UST is applied from 750°C to 660 °C without affecting the onset temperature of α-Mg nucleation. One of the major issues with grain refinement of Mg alloys with Zr addition is the vast difference in the density (~74%) between Mg and Zr particles at alloying temperatures (750 °C) [5,6]. As a result, most of the added Zr particles from the master alloy settle to the bottom of the crucible and approximately 2.33 wt.% of the Mg-Zr master alloy is needed to produce a final alloy composition of 0.7 wt.% Zr [4,6,66]. This severe loss in Zr increases the cost of the alloying process and reduces the grain refinement efficiency in the as-cast condition. Significant reduction in the grain size is obtained only when the addition of Zr is increased above the peritectic composition to 0.8 and 1.0 wt.% in the as-cast condition. After UST, the refinement has been notably improved at lower additions of 0.4 and 0.5 wt.% Zr. With the analysis of composition for dissolved and undissolved particles, it is observed that the UST increases the efficiency of alloying from 30% in the as-cast conditions to 66% by increasing the amount of dissolved Zr solute in the alloy. Reduction in the settling of Zr particles, finer size distribution and its activation by UST maximises the overall efficiency of grain refinement (< 100 µm grain size) for Zr > 0.5 wt.% [6].

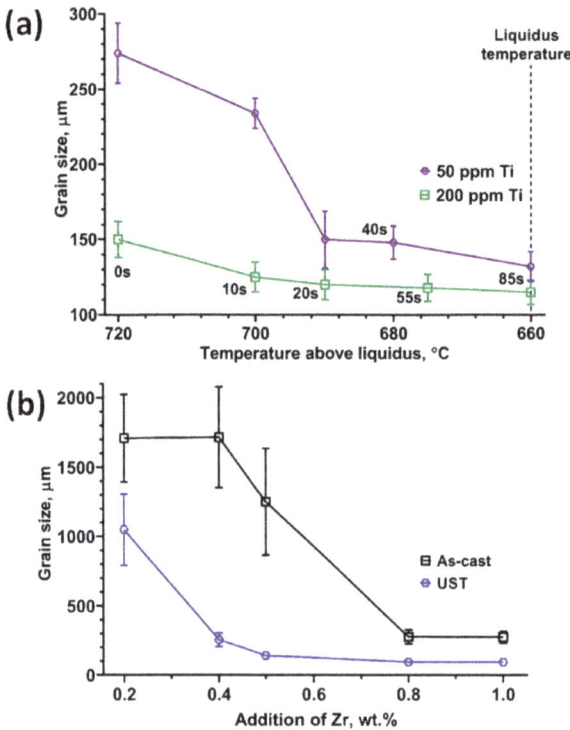

Figure 6. UST terminated above liquidus temperature at specific temperatures from 720 °C to 660 °C in (**a**) Al$_3$Ti1B master alloy added at 50 and 200 ppm to pure Al and (**b**) Mg-25Zr master alloy added to commercial purity Mg. The liquidus temperature in Mg-Zr alloys varies from 651 °C to 653 °C in which UST is applied from 750 °C to 660 °C for 60 to 90 s.

4.3. Refinement of Primary Intermetallic Phases

One of the major reasons for the increase in the total number density of particles is due to the refinement of primary intermetallic particles by UST and thus increases the population of active substrates for grain refinement [6,47]. It has been reported that the narrow size distribution of refiner particles after UST of Zr added to Mg [6] and Al-Ti-B [67], Al-Zr, Al-Zr-Ti [63] added to Al alloys results in better grain refinement. The refinement of blocky and needle shape primary intermetallic phases such as Al$_3$Ti [63,68], Al$_3$(Zr, Ti) [63,69], primary Si [70–72] and β/δ-Al-Fe-Si [24,25] phases is an added advantage of UST that further improves the mechanical properties of the cast alloys. Figure 7 shows a summary of the common intermetallic phases encountered in commercial Al alloys and their refinement after UST. In Al-Si alloys, both the primary and secondary Si phases are refined. From analysis of cooling curves, it was found that the nucleation temperature of primary Si phase increased [70], indicating that more oxide particles are activated at a smaller undercooling for enhanced nucleation of Si. Direct observation using synchrotron studies reveals that streaming flow, cavitation bubbles and flushing of hot fluid to the roots of the primary phase can cause detachment and remelting for the refinement of primary phases [29,30].

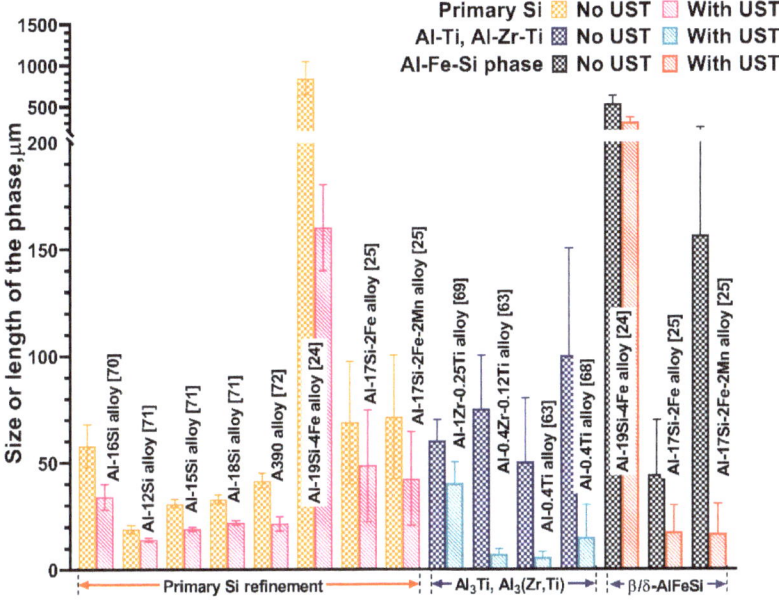

Figure 7. Refinement of primary intermetallic phases after UST reported for selected Al alloys.

Besides refinement in the size of the intermetallic phase, UST also favours the completion of the peritectic reaction reported in Al and Mg alloys [6,24,25]. The sluggish dissolution reaction of Zr in Mg has been improved after UST and increases the solute content to the equilibrium concentration of 0.5 wt.% [6]. Todaro et al. [24] systematically investigated the refinement of primary Si and AlFeSi phases of $Al_{19}Si_4Fe$ alloys with UST. Compared to the as-cast ingot, UST results in the refinement of Fe and primary Si phases, improved their distribution, reduces macro-segregation and large shrinkage cavities. Acoustic streaming produces effective mixing and prevents the settling or floating of intermetallic phases. The peritectic reaction (L + δ-Al_3FeSi_2 → β-Al_5FeSi + Si) in the as-cast condition is incomplete due to the formation of large blocky shape platelets of δ-Al_3FeSi_2 and only fine precipitates are expected to undergo complete transformation. After UST, most of the intermetallic particles contain β-Al_5FeSi and only a few δ-Al_3FeSi_2 complex phases exist, indicating that a complete peritectic transformation is favoured. Similarly, it is found that the addition of Mn combined with UST in $Al_{17}Si_2Fe$(0.5 to 2.0)Mn alloys favoured the transformation to the desirable α-$Al_{15}(Fe,Mn)_3Si_2$ phase finely distributed with polygonal morphology [25]. The refinement of primary phase and its morphology is generally reported to be affected by both nucleation and fragmentation effects depending on the temperature range of UST applied during solidification [24,25,68].

In alloys containing potent particles (either peritectic systems or when external heterogeneous particles are added), UST can be employed either isothermally above the liquidus temperature or during cooling and terminating UST just above the liquidus temperature. In both these conditions grain refinement is improved by:

(i) De-agglomeration of particles by cavitation and simultaneous improvement in activation (wettability) of the particles by the penetration of parent liquid into the surface defects of the particles [2,47,50,73–75];
(ii) Excellent dispersion of the particles by acoustic streaming, as the melt in the liquid state has less resistance to fluid flow [6,47];

(iii) narrow size distribution of nucleant particles throughout the melt and a reduced distance between adjacent potent nucleant particles [6,46,63];
(iv) Reduced loss of inoculants by agglomeration and settling to the bottom of the casting [6];
(v) Fragmentation of large, coarse primary intermetallic phases to fine structures that increase the rate of nucleation of primary grains [24,63,69,70].

5. UST Applied from Above to Below Liquidus Temperature during Solidification

5.1. With the Addition Master Alloys Containing Potent Refiner Particles

The grain refinement produced by UST when applied in the liquid melt or terminated just above the liquidus temperature mainly depends on the heterogeneous particles. For example, potent particles like TiB_2 and Zr produce excellent refinement compared to oxide substrates [6,44,46,47]. On the other hand, when the UST temperature range is extended to below the liquidus temperature, the refinement is enhanced to a greater extent even without potent particles or at reduced amounts of inoculant addition. Figure 8 shows that the grain density is remarkably increased when UST is applied below the liquidus temperature in Al containing TiB_2 particles and the difference in refinement with respect to the amount of Al_3Ti_1B master alloy (50, 200 ppm) is almost insignificant. In other words, the amount of grain refinement obtained with 200 ppm could be readily obtained with 50 ppm of Ti when UST continues to be applied below the liquidus temperature for 100 s. It is interesting to note that the grain density increases rapidly until 20 s and reaches a steady state condition until 80 s. This implies that for a given addition amount of refiner, there is a limit to the number density of the particles that can be activated, in which a further activation is possible by extending the UST temperature range to below the liquidus temperature. This is also observed in a Mg-0.2Zr alloy in which a 63% grain size reduction after terminating UST above liquidus temperature has been further increased to 85% when UST is extended below the liquidus temperature for 2 min. This is a significant advantage of UST where the refiner addition can be reduced without compromising the percentage of refinement obtained [6].

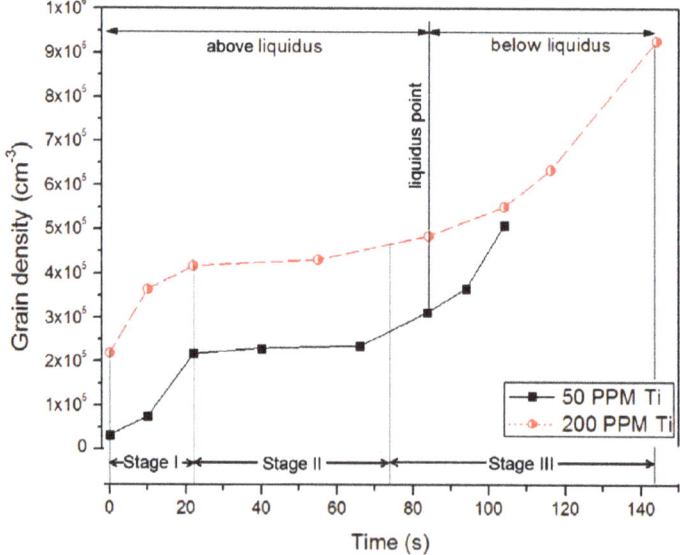

Figure 8. Increase in the grain density with respect to UST time when UST is extended below liquidus temperature of pure Al containing 50 and 200 ppm of Ti [47].

5.2. Pure Metals and Alloys

Without the addition of potent grain refiners, Wang et al. [52] investigated the grain refinement achieved when UST was applied to an Al-2Cu alloy during solidification over different temperature ranges from above the liquidus to complete solidification. It was found that when UST is terminated above liquidus temperature, there is no refinement in the grain size. However, a significant refinement in grain size (150–200 µm) is noted when UST is applied from 40 to 60 °C above the liquidus temperature and continued below the liquidus temperature for 4 min. Reducing the starting temperature of UST to 20 °C above the liquidus or less than that results in the formation of coarse grain structure similar to the as-cast grain structure due to the formation of a solid chill layer beneath the unpreheated sonotrode. With the use of a preheated sonotrode (heated to 285 °C) the formation of a chill layer is avoided, and acoustic streaming was established in the melt to transport grains to form an equiaxed grain structure [38].

To understand the mechanism of grain refinement without the interaction of solute or potent particles, pure metals (Mg and Zn) were investigated over different time and temperature ranges. Figure 9 shows the macrostructure refinement after UST was applied to Mg and Zn. Superheat refers to the starting temperature of UST above the melting/liquidus temperature of the alloy. A high superheat temperature (UST turned on at 100 °C above the melting temperature) terminated after 3 min (until complete solidification) produces a coarse and non-uniform grain morphology in Mg (Figure 9a) [44]. When the superheat temperature range is reduced to 40 °C above the melting temperature, the grain refinement is homogeneous and a completely refined structure is obtained throughout the casting with a more uniform distribution of grains (Figure 9c). Interesting insights were revealed from Zn solidification. UST applied from 30 °C above melting temperature for 4 min results in refinement from the as-cast condition, however, the grain structure is completely dendritic in the centre of the casting (Figure 9b). Under similar conditions, a low superheat of 20 °C for 3 min produces an equiaxed zone with non-dendritic grains in nearly half of the casting's cross section (Figure 9d) [43]. In the as-cast condition, both these metals exhibit a grain size range from 2 to 3 mm on average [43,44,46]. After UST the grain size (in the equiaxed zone) ranges from 160 to 400 µm which is more than 90% reduction in grain size from the respective as-cast conditions. Figure 9e shows a plot of the equiaxed grain area measured from the cross-section area of the casting in pure Mg, Zn and Al-2Cu alloy with respect to UST time [43,44,76]. At a given low-superheat temperature range, increase in the area fraction of equiaxed grains is proportional to the time of UST applied below the liquidus or melting temperature. Depending on the thermal properties of the metals, approximately 1 to 2 min during solidification is enough for Al and Mg to achieve an equiaxed zone throughout the cross section compared to Zn.

Qian et al. [40] proposed that (in magnesium alloys) these grains are nucleated during UST as a result of cavitation and dispersed into the melt by acoustic streaming. By measuring the grain size with respect to distance from the sonotrode, it is found that the fine grains are observed closer to the sonotrode and the grain size increases with distance from the sonotrode to the bottom of the casting. Therefore, these grains are assumed to nucleate in the cavitation zone beneath the sonotrode and then dispersed into the melt. Due to the attenuation of sound waves, the grain size becomes coarser in the bottom region of the casting near the crucible wall. As grain refinement is observed only when the UST is applied during solidification includes the onset of nucleation, some researchers believe that fragmentation of dendrites by cavitation was the major reason for the grain refinement rather than independent nucleation [2]. The mechanistic viewpoints on grain formation will be detailed in the later sections, however, the grain refinement achieved by extending UST below liquidus temperature is promising because it eliminates or reduces the need for the addition of external particles.

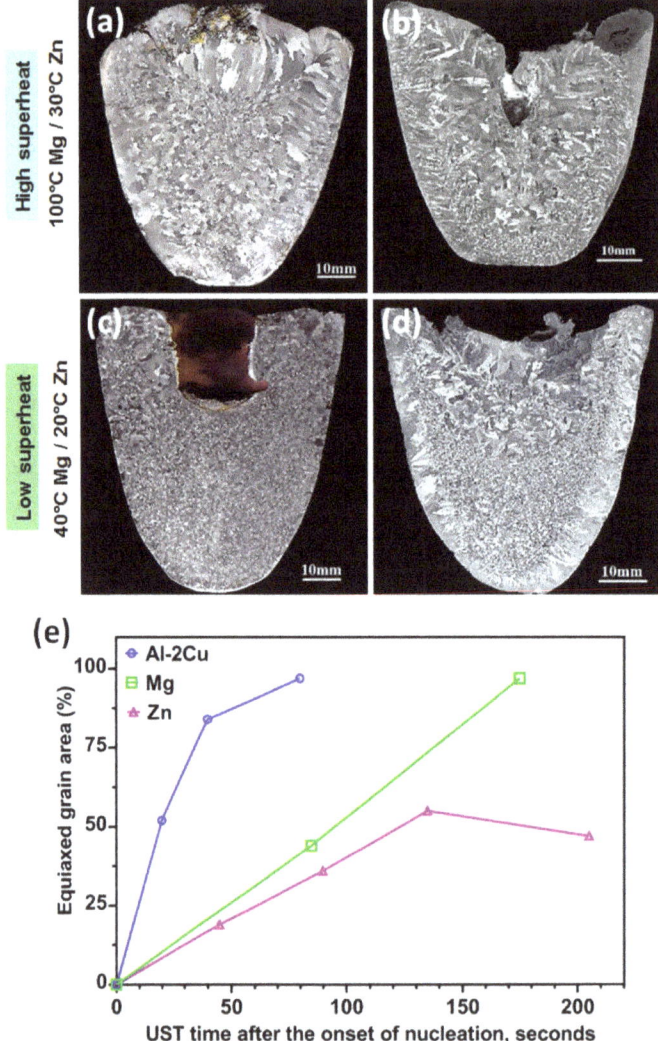

Figure 9. Macrostructures after UST was applied during solidification at (**a**,**b**) high and (**c**,**d**) low superheat (temperature at which UST is turned on above the melting temperature) in (**a**,**c**) pure Mg and (**b**,**d**) pure Zn [43,44]. (**e**) Equiaxed grain area measured from the cross section of the casting with respect to UST time after the onset of solidification [43,44,76].

The additional factors (temperature range, time duration and alloying elements) that contribute to deliver grain refinement when UST is applied below the melting or liquidus temperature in alloys are:

(i) Increased number of nucleation events on heterogeneous potent particles compared to UST terminated above liquidus temperature [6,44,47,53];

(ii) Reduction in the temperature gradient of the bulk liquid under the action of acoustic streaming promoting nucleation on potent particles and assists the survival of grains [45,52,57,77];

(iii) Formation of fine non-dendritic grains below the sonotrode at the liquid–sonotrode interface due to the colder vibrating source and then distributed by acoustic streaming into an undercooled melt [6,20,40,43,46];

(iv) Fragmentation of dendrites caused by the interaction of cavitation and acoustic streaming in the mushy zone or at the solid–liquid interface [32,34,35,41,78].

6. Grain Formation Mechanisms and Development of the Refined Ingot Structure

6.1. Origin of Equiaxed Grains

The dominant mechanism of UST grain refinement is often debated between cavitation causing fragmentation of dendrites [26,27,32,42] and an enhanced nucleation mechanism based on the heterogeneous substrates [1,40,74,79]. Several experiments under a high intensity X-ray synchrotron technique have shown that fragmentation or fracturing of primary phases by cavitation bubbles are responsible for the refinement [34,41,42,78]. However, alloys treated under UST in a crucible exposed to room temperature supports an enhanced nucleation mechanism, where the cavitation induced pressure pulses are expected to increase the rate of nucleation on heterogeneous substrates. The grains produced in this condition show more spherical and non-dendritic morphologies that are much finer than the secondary dendritic arm spacing of the alloys without UST. Therefore, the fragmentation of well-developed grains in this case cannot be the cause of the formation of a fine non-dendritic grain structure. If these grains are expected to be generated by early stage fragmentation then the size of such crystals and its survival rates are questionable [74].

To better understand the mechanism of the origin of these non-dendritic grains two approaches were followed to characterize grain formation in the cavitation zone (i) placing a gauze encapsulating the cavitation zone and (ii) using quartz tubes to extract the melt during UST and quenching immediately [76]. Using the gauze setup it was found that the cavitation zone immediately beneath the sonotrode is responsible for grain formation. Figure 10a clearly shows that non-dendritic, fine grains are directly observed at the sonotrode-liquid interface region of the casting. The areas separated by the gauze only show a coarse grain structure (similar to as-cast structure) including adjacent to the mould walls and the top surface of the casting. This confirms that the non-dendritic grains originate in the cavitation zone beneath the sonotrode.

Figure 10b shows the points in the cooling curve (1 to 6) where tube samples were taken from the solidifying melt and the microstructures from selected sampling stages (1, 3, 4 and 6). As UST has already started at 40 °C above the liquidus temperature, grains are produced beneath the sonotrode when the temperature reaches the liquidus temperature and then transported into the liquid melt. The microstructure of sample 1 taken at the liquidus temperature shows large dendritic grains formed during quenching of the sample, meaning that the number density of non-dendritic grains is very low. Some non-dendritic grains started to appear in the sample 2 at 10 s after the onset of solidification and continue to increase up to sample 6, at 80 s where the microstructure of the entire tube sample has a large number of non-dendritic grains. Therefore, the grains generated at the start (<20 s) were pushed down by the acoustic streaming force leaving a lower number density of spherical grains at the top of the melt. After 20 s, both the continuous formation of new grains below the sonotrode and the recirculation of the existing grains starts to fill the casting cross section to nearly half of the volume (refer to Figure 11a) and more fine grains start to appear in the top region after 80 s.

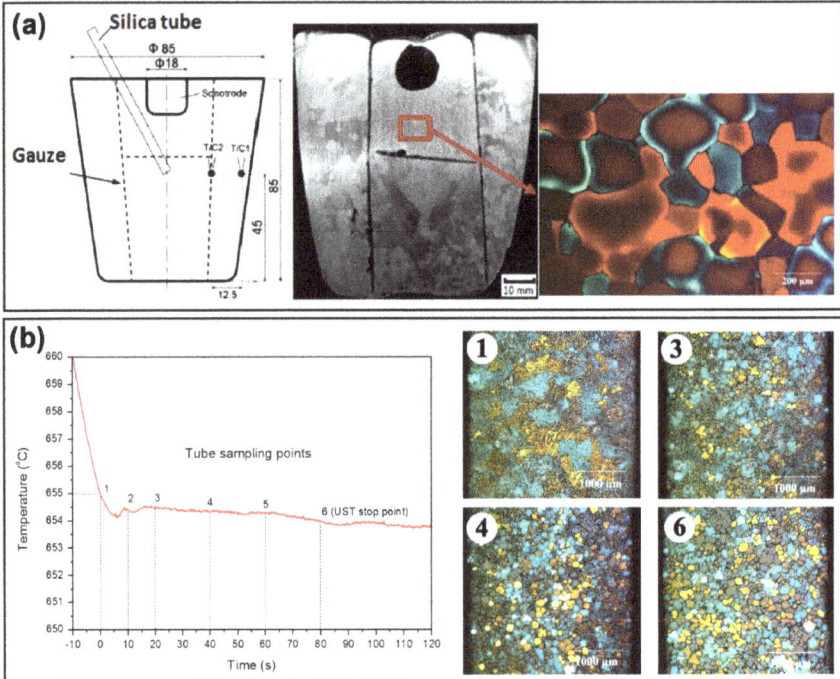

Figure 10. (**a**) Schematic of the UST setup that uses a stainless-steel mesh to capture the grains beneath the molybdenum sonotrode and the resultant macrostructure and microstructure (taken from the gauze area). (**b**) Cooling curve of Al-2Cu alloy with numbers denoting the time at which tube samples were taken during UST and the corresponding microstructures of samples 1, 3, 4 and 6 [76].

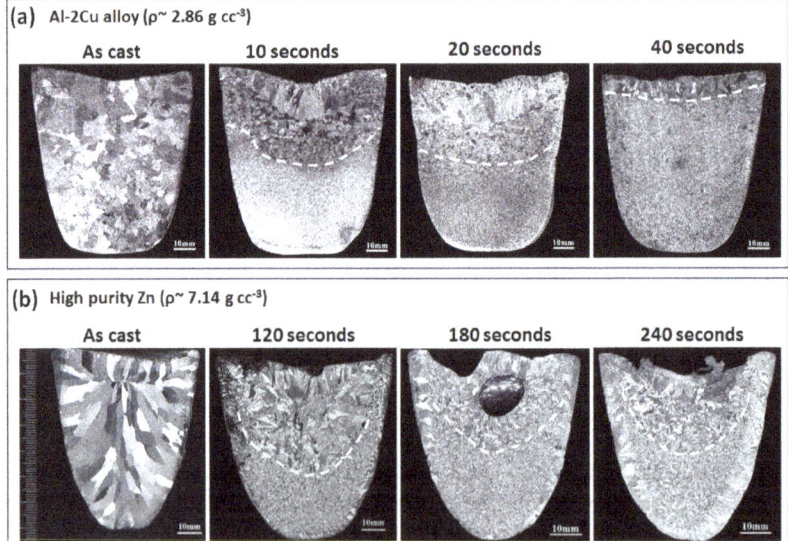

Figure 11. Settling of grains when UST is terminated after various time periods during solidification of (**a**) Al-2Cu alloy and (**b**) high purity Zn [43,76].

6.2. Settling of Grains, Effect of Volume and Height of the Casting

When UST is applied until complete solidification, the whole ingot structure is refined as shown in Figure 9c. As already described in Figure 10b increasing the time gradually increases the number of non-dendritic grains. Figure 11 shows the effect of such grains settling towards the bottom of the crucible when UST is terminated before complete solidification. After the grains are generated in the cavitation zone, the acoustic stream carries these grains to the bottom of the casting [43,76]. Low temperature gradients enhance the survival of a higher number density of grains and it results in the non-dendritic equiaxed zone. Termination of UST at shorter times results in two types of grains above the equiaxed zone (i) rosette or mixed dendritic grains just above the equiaxed layer and (ii) large columnar grains at the top surface of the casting influenced by the radiation heat transfer [76]. Based on the thermal conductivity of the metal, the action of grains filling the macrostructure of the Al-2Cu alloy ingot (Figure 11a) is faster (80 s) than that of pure Zn (240 s, Figure 11b). Furthermore, it is interesting to note that increasing the time from 20 to 80 s (Al-2Cu alloy) produces completely equiaxed grains throughout the cross section of casting whereas in Zn increasing the time duration from 180 s to 240 s results in a similar area of equiaxed zone. Also, the temperature range at which excellent refinement is achieved is 40 °C above the liquidus temperature in Al-2Cu alloy and for Zn it is 20 °C above the melting temperature. This comparison shows that UST produces similar tendencies in grain formation and settling regardless of the type of metal and it only depends on the temperature range of UST.

The role of acoustic streaming in transporting the grains for large volumes varying from 137 to 946 cc is shown in Figure 12. The increase in the crucible size led to an increase in the lateral volume in castings 1 to 3 and simultaneously increases the height from 6.5 cm for casting 1 to 12 cm for casting 3. Casting number 4 has the longest distance of 17.5 cm from the sonotrode tip to the bottom of the casting. In all these cases, the sonotrode was immersed into the melt from the top to nearly 1–1.5 cm below the surface. For all the heights of the castings investigated, the application of UST from 40 °C above the liquidus until complete solidification produces refinement throughout the casting's cross section (castings 1 to 4). However, macro examination of castings 3 and 4 reveals that the grains are slightly coarser than castings 1 and 2. The grain size measured from the top and bottom of castings 1 to 4 is shown in Figure 12b. As the volume and distance increases the grain size steadily increases in the bottom region. For larger castings (2, 3, 4) the grain size in the cavitation zone also shows bigger grains with large deviations compared to casting number 1. In castings 3 and 4, mostly mixed grain structures (grains with rosette and dendritic morphology) are observed throughout the ingot with few non-dendritic grains. As UST is initiated well above the liquidus temperature of the alloy, it is possible that these grains while moving towards the bottom of the crucible, would grow to take a rosette or dendritic form. Increasing the height of the casting above the distance of $\lambda/2$ started to show a fading tendency in the degree of grain refinement. Nonetheless, the refinement after UST is significant compared to the respective as-cast conditions.

6.3. Evolution of Grain Structure, Morphology of Grains and the Role of Alloying Elements

Depending on the properties of the metal (melting temperature, density and thermal conductivity) significant differences were noted in the UST grain structures of pure Zn, Al and Mg. Figure 13 shows the macrostructure of pure Al and Zn where UST is applied from 40 °C and 30 °C above the melting temperature, respectively, until complete solidification.

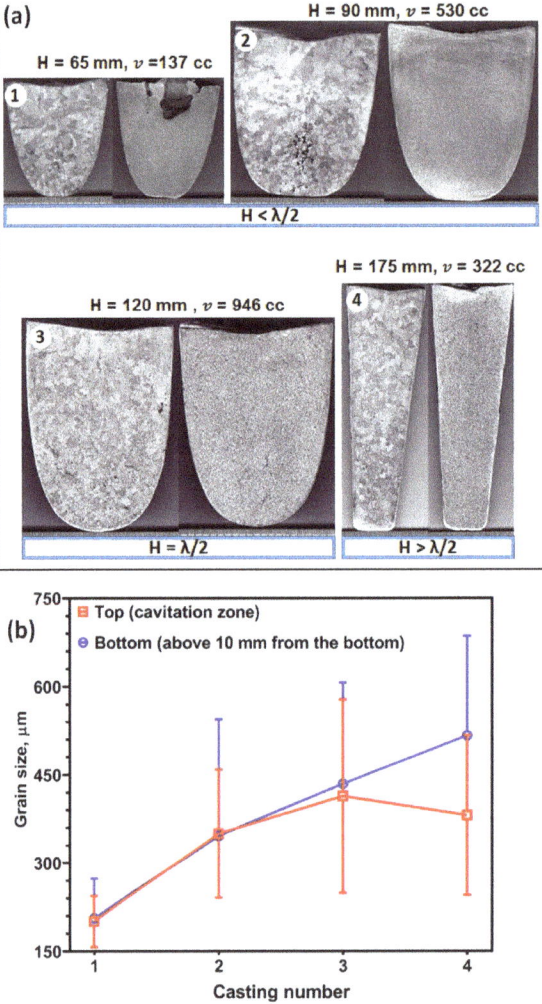

Figure 12. (a) Effect of casting volume (*v*) and height (H) on the macrostructure after UST of an Al-2Cu alloy and (**b**) the grain size measured from the top and bottom of the casting. (λ/2 = 125 mm refers to the half wavelength distance of the sonotorde).

The grain structure in the Al ingot is uniform with only equiaxed grains throughout the cross-section. Pure Mg solidified under similar conditions also shows a fully equiaxed structure for the whole cross section in Figure 9c. On the other hand, both columnar and equiaxed grains were found after Zn solidification under UST. From the macrostructures shown in Figure 9d, the Zn columnar grains grew from ~4.0 mm to 10.6 ± 0.6 mm. According to the grain formation mechanism explained in the Figure 10a these grains filled the small volume of Mg and Al alloys within shorter durations. Due to the larger solidification interval and low melting temperature of pure Zn, any grains created above the melting temperature have a greater chance of remelting. While applying UST below the melting temperature, acoustic streaming reduces the steep temperature gradients in the centre of the melt and the relatively colder zones of the mould started to nucleate columnar grains. These columnar grains have finer width (~0.7 mm) and are numerous along the mould wall compared to the as-cast condition of pure Zn (~3.0 mm) [43]. The increase in the length of the columnar grains growing perpendicular to the direction of the sonotrode indicates that (i) the temperature ahead of the melt is lowered by the acoustic streaming and (ii) there is no obstruction to the continued growth of the columnar grains by the circulating grains. Therefore, fine grains that are formed during this condition are expected be lower in number density and tend to settle quickly towards the bottom of the crucible, allowing the columnar grains to grow from the side wall without impingement.

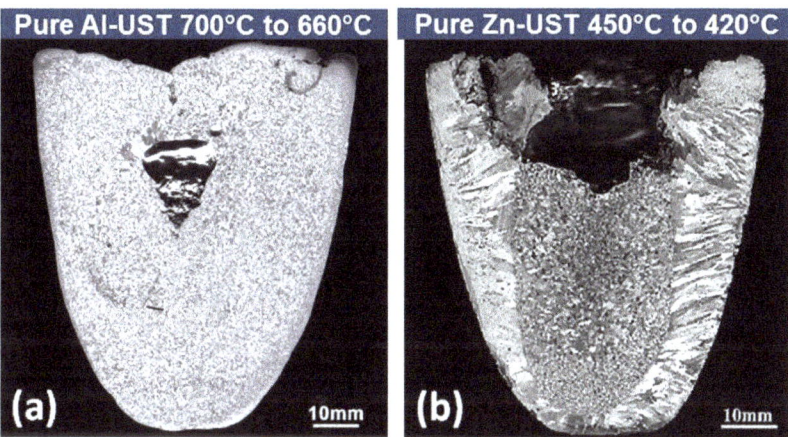

Figure 13. Macrostructures of ultrasonically treated (**a**) pure Al and (**b**) pure Zn.

Comparison of grain structures with respect to the temperature range of UST for pure Zn is shown in the Figure 14. The cooling curve in Figure 14 shows three ranges of UST A, B and C. An unpreheated titanium sonotrode was used in all these experiments. The microstructures from each casting condition taken from the centre of the casting is shown in Figure 14A–C.

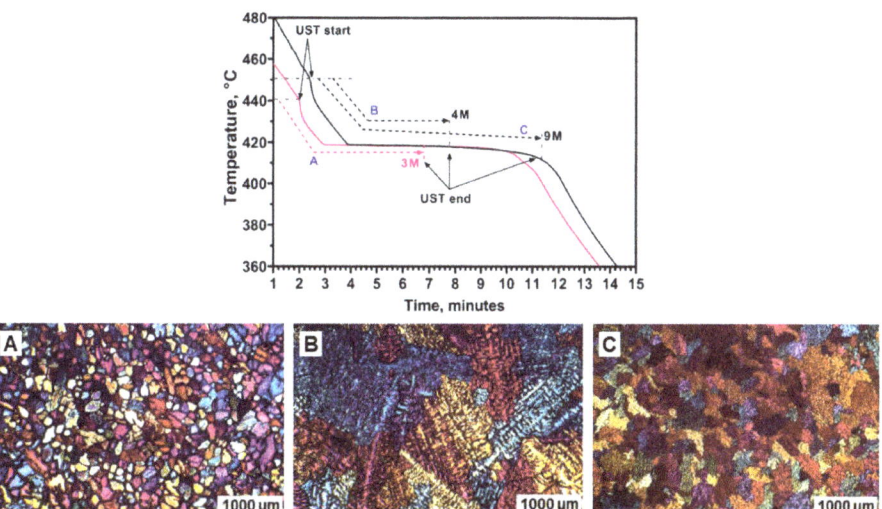

Figure 14. Cooling curve of pure Zn showing three different temperature ranges of UST application **A**, **B**, **C** and their corresponding microstructures taken from the centre of the casting.

The low temperature range of UST for a shorter time (440 °C-3 min) results in non-dendritic grains in microstructure A. Increasing the temperature range to 450 °C for a longer time (4 or 9 min) results in either coarse dendrites (B) or equiaxed grains with dendritic morphology (C). It should be noted that the equiaxed grains in A are completely non-dendritic whereas the grains in C are dendritic equiaxed. 3 min of UST produces non-dendritic grains in nearly half of the cross section, because of the low superheat temperature range (A). However, 9 min UST at a slightly higher temperature range produces only dendritic grains (C) in the equiaxed zone and promotes columnar grain growth from the mould wall (Figure 13b). These grain structures show that the formation of non-dendritic grains is related to the low-superheat temperature range of UST. When a higher starting temperature is used, the sonotrode is heated to a higher temperature and results only in dendritic grains. Such clear observations noted in pure Zn are not found in pure Mg or Al and its alloys during UST solidification.

To further understand the grain formation effect of an unpreheated sonotrode, UST is applied to pure metals just before the completion of solidification in the equilibrium melt as shown in Figure 15.

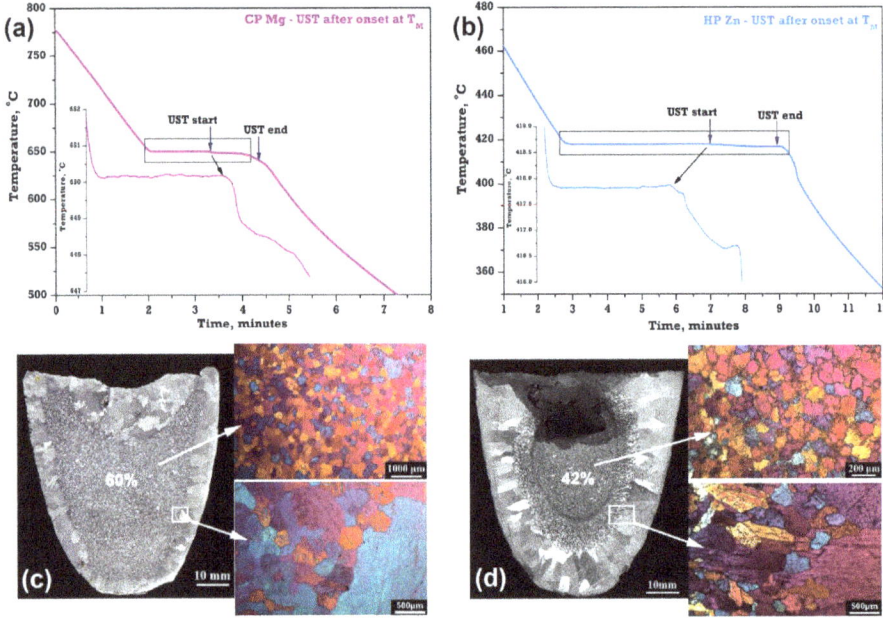

Figure 15. (**a,b**) Cooling curves, (**c,d**) grain structures in the macrostructure and microstructure of pure (**a,c**) Mg and (**b,d**) Zn.

The cooling curves of Mg and Zn in Figure 15a,b show that UST is applied only at the end of complete solidification without affecting the onset of nucleation. Thermocouples were placed slightly offset to the sonotrode (Figure 15a,b) and it was found that there is a significant drop in temperature of the pure metal (~1 to 1.5 °C). As explained in the Figure 10, the grains generated below the sonotrode are pushed downwards into the equilibrium melt to create equiaxed grains in nearly 40% to 60% of the cross section within 1 to 2 min of UST. Microstructures observed in the centre of the casting shows that these grains in both Mg and Zn castings were non-dendritic (Figure 15c,d). The forced downward movement of the fine grains due to acoustic streaming impinge on the columnar grains to form the columnar to equiaxed transition. Therefore, considering the above discussion and the solidification conditions, fragmentation of existing dendrites is less likely to be a significant contributor to the refinement.

Figure 16 shows the effect of important alloying elements reported in Al and Mg alloys for grain refinement after UST under similar casting conditions. Incremental additions of solute Mg [49], Cu, Ni [79] and Si (< 4 wt.%) [45] to Al alloys produces an average grain size less than 400 µm after UST. When grain refiners are present (TiB$_2$ [46,47,53] and Zr particles [1,6]) the effectiveness is further improved even at low additions. It is well-known that the addition of Si > 3–4 wt.% increases the grain size in Al alloys, where the addition of Al-Ti-B refiners cannot produce significant refinement in the as-cast condition [18,19]. The grain coarsening behaviour of Al alloys containing Si without and with Ti is shown by solid lines in Figure 16. UST, on the other hand, refines the grain size in both these alloys (Al-Si and Al-Si-Ti [45]) at the temperature range of 40 °C above the liquidus temperature to complete solidification. Research on Mg alloys (containing Al [80] and Zn [3]) has also shown similar results for UST refinement as a function of solute concentration.

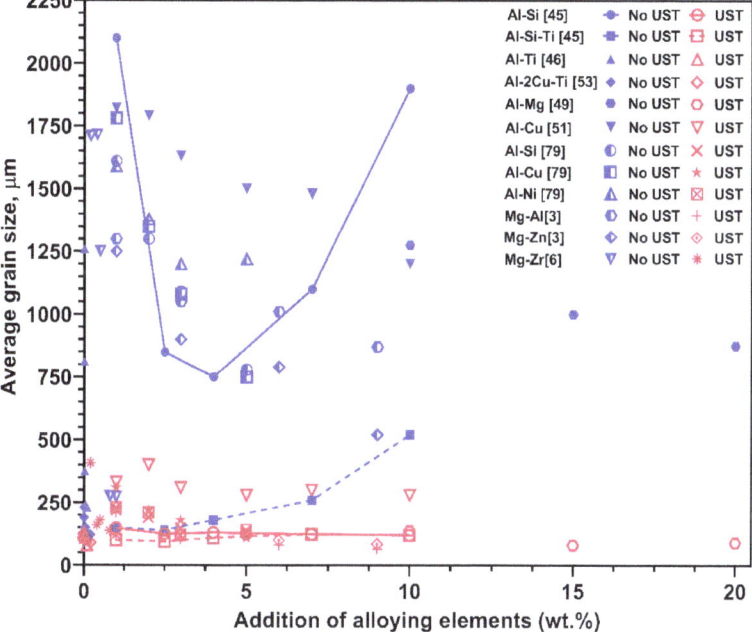

Figure 16. Role of alloying elements on the grain refinement achieved when UST is applied during continuous cooling from above to below the liquidus temperature.

7. Interpretation and Application of the Interdependence Model for Solidification under UST

The Interdependence model [15] is a useful framework for analysing the factors that can be controlled in order to optimise an alloy's grain size (d_{gs}) by facilitating nucleation. The model is described by Equation (4).

$$d_{gs} = \frac{D \cdot z \Delta T_{n-min}}{vQ} + \frac{4.6D}{v} \cdot \left(\frac{C_l^* - C_0}{C_l^* \cdot (1-k)} \right) + x_{Sd} \qquad (4)$$

where D is the solute diffusion coefficient, ΔT_{n-min} is the nucleation undercooling required to nucleate on the most potent particle, z is the incremental amount of ΔT_n that needs to be generated by constitutional supercooling (ΔT_{CS}) for a subsequent nucleation event to occur, v is the growth rate of the grain–liquid interface, C_o is the alloy composition, C_l^* is the composition of the liquid at the interface and k is the partition coefficient. The three terms calculate the elements that make up the grain size such that:

$$d_{gs} = x_{CS} + x'_{dl} + x_{sd} \qquad (5)$$

where x_{CS} is the growth of previous grain to generate $\Delta T_{CS} = \Delta T_n$, x'_{dl} is the length of the diffusion field to where ΔT_n is achieved, and x_{sd} is the average distance to next most potent particle. The first two terms $(x_{CS} + x'_{dl})$ represent the nucleation free zone x_{NFZ}. Note that x'_{dl} is controlled by x_{CS} that establishes the value of C_l^* in Equation (4). In order to reduce the grain size either, or both, of x_{NFZ} and x_{sd} need to be decreased.

The following discusses how these factors are relevant to solidification under UST. From the results described in the previous sections, Q is clearly important and can be readily manipulated. When Q becomes very large x_{NFZ} will tend to zero such that the number of nucleation events corresponds to x_{sd} thus the particle number density controls the amount of nucleation. If the nucleant particles have a very high potency tending towards epitaxial nucleation then ΔT_n tends to zero and thus x_{NFZ} tends to zero. This effect has been demonstrated by the addition of niobium boride particles to an Al-Si alloy where the effect of Si poisoning is eliminated because the niobium boride particles are large and have a very good orientation relationship with aluminium thereby reducing x_{NFZ} to a small value [81]. The interfacial growth rate v will be relatively slow as the cooling rate is low and growth occurs near the liquidus temperature. The solute diffusion rate in the liquid, D, may be enhanced by convection associated with acoustic streaming. However, acoustic streaming has a much more significant effect on the temperature gradient as discussed in Section 3. Thus, the term z in Equation (1) tends to zero because the temperature gradient becomes flatter.

Considering the factors that can be controlled the most important are Q by adding growth restricting elements to the alloy, reducing z through acoustic streaming, and reducing ΔT_n by adding potent nucleant particles. Also, x_{sd} can be reduced by increasing the particle number density of these potent nucleants.

The results from the UST studies conform to the expectations of the Interdependence model. However, this is surprising since the majority of nucleation events occur in the cavitation zone directly under the sonotrode and not in the bulk of the melt. So why is the Interdependence model effective in predicting significant refinement when UST is applied? The answer lies in consideration of the role of acoustic streaming. Because acoustic streaming flattens the temperature gradient in the bulk of the melt, the melt cools with essentially the same amount of undercooling throughout the casting. Therefore, when the grains formed in the cavitation zone are swept into the melt they move into a melt that is undercooled. Also, for alloys, the solute rejected during grain growth creates a constitutionally supercooled layer which protects the grains from remelting [82]. The higher the value of Q the faster CS is generated providing greater protection from remelting. Therefore, the combination of a low temperature gradient and high Q are critical for the survival of grains leading to a finer grain size. If this situation is satisfied the next biggest effect is the addition of a high grain number density of potent particles to reduce x_{sd}.

x_{sd} has a different meaning for the UST conditions used in our experiments. Because nucleation of equiaxed grains occurs under the sonotrode and not in the bulk of the melt, x_{sd} is defined by the grain number density and not by the number density of potent particles that are able to be activated. Based on the study of grain formation, just below the sonotrode the grain density increases with time during UST as shown in Figure 17. However, in the bulk melt the grain number density initially increases quickly due to fewer small grains with plenty of room to move and grow eventually reaching a maximum after about 30 s when the density becomes higher where grain to grain interactions in the melt become common. After 30 s the number of grains keep increasing while the size of the equiaxed zone increases as shown in Figure 11, but the grain number density does not change significantly. This means that the value of x_{sd} changes during solidification and it is difficult to predict these changes in a casting as x_{sd} also decreases due to the settling of grains towards the bottom of the casting and increases in the top region of the casting due to the depletion of grains.

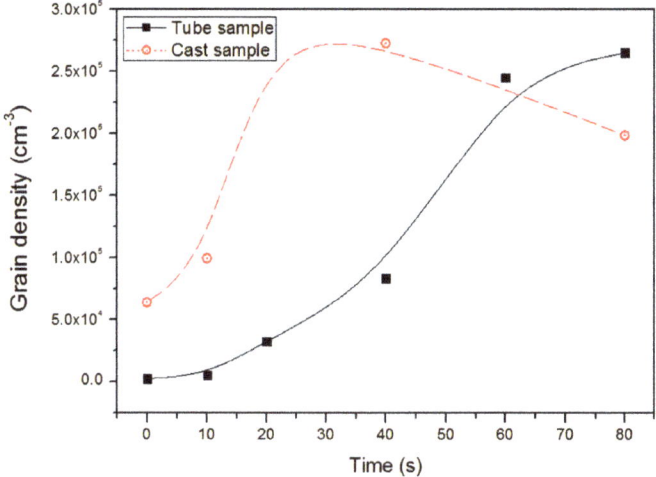

Figure 17. Comparison of grain density measurement of castings (Figure 11a in [76]) and tube samples (Figure 10b in [76]).

In plots of grain size versus $1/Q$ (Figure 18a,b) x_{sd} corresponds to the final grain number density after UST was terminated and settling has finished. Because x_{sd} under UST conditions is based on grains rather than particles the difference with changes in alloy composition are relatively small. Therefore, the role of settling is very important in controlling the grain size as highlighted by Figure 12a. Adapting the Interdependence model to take settling into account is a challenge due to density differences between liquid and grains (e.g., the density differences over a range of Al-Cu compositions) can change dramatically from promoting settling to resulting in floating of grains [83], and between different alloy systems. Despite this difficulty the Interdependence model is still a useful tool for determining the mechanisms controlling grain size.

The following three examples used the Interdependence model to determine the mechanisms responsible for the grain sizes achieved. Figure 18a shows a plot of grain size versus $1/Q$ for eutectic systems reported for Al and Mg alloys. As these alloys have no active potent substrates for nucleation in the as-cast condition, grain sizes were larger. UST was applied to these alloys from above to below the liquidus temperature during solidification. After UST the grain size was significantly reduced (<500 µm) in low solute containing alloys and for Q values exceeding 10 K the grain refinement becomes excellent (<100 µm). The shaded region between the as-cast and UST refined alloys is x_{NFZ} where x_{NFZ-1} and x_{NFZ-2} highlight the difference between the dilute and high solute alloys, where dilute alloys show grain sizes in the range of 200 to 400 µm. It is interesting to note that grain refinement in conventional casting conditions is largely dependent on Q values with a steeper slope, however, after UST the trend of refinement appears to be much flatter regardless of the type of eutectic forming solutes. Figure 18b shows the effect of Zr solute and particles in Mg and Ti solute and TiB$_2$ in Al alloys after UST. Due to the potency of particles and higher Q values, most of the data points in the as-cast condition fall into the significant refinement range (<100 µm), except at very low additions. An increase in the intercept is noted for Al-Si alloys containing Ti due to Si's poisoning effect, however, UST produces excellent refinement of those alloys.

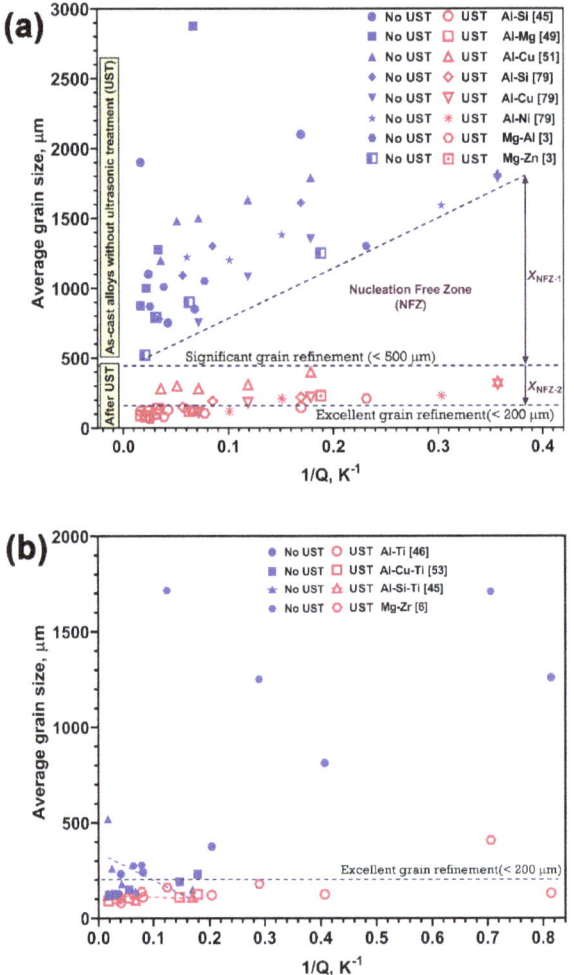

Figure 18. Grain size vs. 1/Q graphs for (**a**) eutectic and (**b**) peritectic alloys solidified under UST.

During UST, the possibility of the formation of new grains is increased rapidly in the cavitation zone [43,44,76] and also by cavitation as a result of physical fragmentation effects [32,33,41,42]. In alloy systems containing potent particles such as Mg-Zr alloys that solidify as equiaxed grains in the as-cast condition, it is assumed that UST preferentially activates nucleation on the Zr particles rather than fragmentation [6,82]. Using the number density of the particles (N_v) and the weight fraction of particles (w_p) estimated through chemical analysis, Equation (4) was modified to:

$$d_{gs} = \frac{1000}{\sqrt[3]{w_p \cdot N_v}} + \frac{D}{v}\left[\frac{0.719 \cdot z}{d_p \cdot Q} + \frac{4.6\left(C_l^* - C_o\right)}{C_l^* \cdot (1-k)}\right] \quad (6)$$

where d_p is the diameter of the Zr particles. Using Equation (6) the predicted grain size versus 1/Q slope of Mg-Zr alloys after UST is found to be in good agreement with the experimental values

For the AlTiB master alloy addition to Al alloys, incorporating the effect of Ti solute and TiB$_2$ particles, a simplified form of Equation (4) can be expanded as [46]:

$$d_{gs} = \frac{1000}{\sqrt[3]{(w_p)_{Al3TiB} \cdot N_{vm}}} + 5.6 \left[\frac{D \cdot z \cdot \Delta T_n}{v \cdot Q} \right] = \frac{32}{\sqrt[3]{(w_p)_{Al3TiB}}} + \frac{652}{Q} \quad (7)$$

Here the weight percentage w_p (TiB_{2A}/TiB_{2MA}) is the ratio of the actual amount of TiB$_2$ added to the alloy to weight percent of TiB$_2$ present in the master alloy and N_{vm} is the number density of TiB$_2$ particles in the master alloy. The slope and intercept values in Equation (7) have been previously quantified for a wide range of Al alloys after TiB$_2$ addition. Figure 19 shows that the effect of the predicted grain size values with the UST processed Al and Al-2Cu alloys after the addition of Al$_3$Ti$_{11}$B master alloy [46].

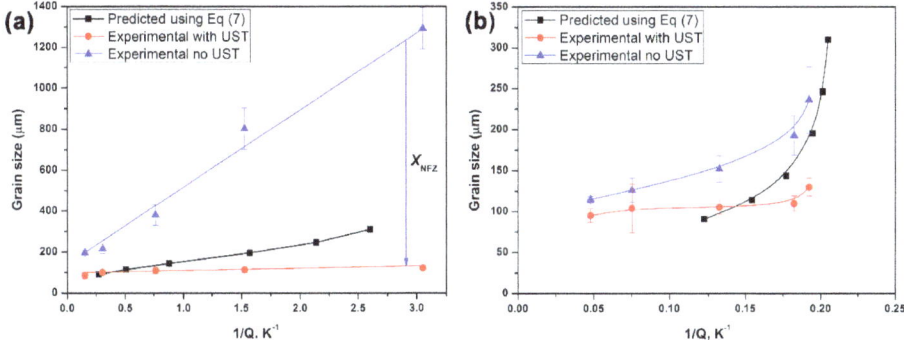

Figure 19. Prediction of grain size using Equation (4) for (**a**) Al-Ti and (**b**) Al-2Cu-Ti alloys with and without UST [46].

The grain size values predicted from Equation (7) have low values of the slope because it assumes that all the added particles were active, however, the experimental result for pure Al (Figure 19a) shows a steeper slope (larger grain size) indicating the x_{NFZ} is larger in the as-cast condition. The prediction trend in Figure 19b for Al-2Cu alloy lies close to the experimental curve, which indicates that the presence of solute Cu facilitates the activation of more TiB$_2$ particles to reduce the x_{NFZ} in the as-cast condition. The grain refinement observed after UST is much finer than the grain size predicted using Equation (7) indicating that more particles are activated to effectively reduce the x_{NFZ} even at low Q values. An intersection of the predicted curves with UST refinement indicates the maximum refinement condition, above which the refiner and UST produces a similar degree of refinement in Ti added Al alloys.

In the above two cases (Zr in Mg and TiB$_2$ particles in Al alloys) the grain refinement after UST was likely to be affected by the potent particles in the range of 0.2 to 2.5 µm. Research by Dieringa et al. [50] showed that the addition of AlN nano-particles with a size range of 20 to 160 nm (with a mean size of 80 nm) to an AM60 Mg alloy produced excellent refinement with a grain size of 85 µm after UST in the liquid only. Unlike the case of UST during solidification below liquidus temperature where acoustic steaming affects the temperature gradient and the nucleation undercooling of particles, the grain refinement in AM60B-1%AlN composites after UST is terminated and then poured into a mould, is only affected by the distribution of particles and constitutional supercooling. By assuming that the largest particles (in this case it's 162 nm with a spacing of about 2 µm) are more likely to nucleate grains, and that x_{NFZ} for constant Q is about 85 µm whereas Equation (4) predicts x_{NFZ} would be larger than 600 µm, suggests that x_{NFZ} is affected by D and/or v according to Equation (8).

$$x_{nfz} = \frac{D}{v}\left[\frac{z \cdot \Delta T_n}{Q} + \frac{4.6\,(C_l^* - C_o)}{C_l^* \cdot (1-k)}\right] \tag{8}$$

Since, the parameters inside the brackets in Equation (8) are almost constant, the reduction of x_{NFZ} in this case is expected as a result of change in either D or v. With the measured undercooling of 14 K after addition of the AlN particles [50], a growth rate of approximately seven times faster than the conventional rate would be needed to achieve the grain size of 85 µm. However, no data is available to support this mechanism. As the diffusion field contains rejected solute and a high number density of AlN nano-particles, solute diffusion will be affected by the presence of nano-particles [84,85]. After reducing the value of D from 5×10^{-9} m²·s to 7×10^{-10} m²·s in the Equation (8), x_{NFZ} predicts the measured grain size after UST of this alloy suggesting that the change in diffusion coefficient could dramatically influence the grain structure.

8. Summary and New Insights

From the analysis of the results of this research and for the casting conditions applied in our experiments, the following provides a generalised description of the key mechanisms affecting the formation of refined grains under UST conditions.

(i) Little refinement is obtained for pure metals and eutectic alloys when UST is applied in the liquid melt. When inoculants or nanoparticles are added to the melt or primary intermetallic and peritectic phases form UST distributes the particles uniformly throughout the melt, breaks up agglomerates of particles and enhances the wetting of particles by the melt. These benefits are realised over a certain UST time for a constant refiner addition, after which no further improvement is obtained. The subsequent solidification occurs as in normal casting processes and the level of refinement achieved can be indicated by the Interdependence model. Application of UST in this temperature range can be readily implemented as part of a foundry's melt treatment process.

(ii) Nucleation of equiaxed grains occurs directly under the sonotrode when UST is applied below the melting point or liquidus temperature. The strong ultrasonically-induced convection transports these grains into the bulk melt while UST continues to produce new grains. The low temperature gradient generated by the acoustic streaming undercools the bulk of the liquid. Refinement can be obtained for pure metals as well as alloys. Alloys with growth restricting solute generate constitutional supercooling around the growing grains providing good protection from remelting further enhancing the level of refinement. The addition of particle inoculants such as TiB_2 and Zr, provides an increase in the number of nucleation events particularly in alloys with a low value of Q.

(iii) The grains formed under the sonotrode are produced at approximately a constant rate implying that the number of grains in the bulk melt are also increasing at about the same rate. The size of the equiaxed region continues to increase until there are enough grains to fill the casting cavity. This process takes about 80 s for Al alloys and over 150 s for Mg and Zn alloys when cast in the standard size ingots.

(iv) Terminating UST at shorter times shows that the suspended grains sink settling on the bottom of the casting after termination occurs. The settling impedes the growth of adjacent grains and columnar grains growing from the mould wall as they pack together. However, for the same UST conditions, when the ingot height is increased the strength of convection decreases with distance from the sonotrode tip due to attenuation and continued settling, at possibly a slower rate, is due to gravity. Thus, the settling grains have more time to grow so that the final grain size is larger. However, refinement still occurs but not to the same extent as in the smaller castings.

(v) UST diminishes the difference due to variation in alloy composition (i.e., Q values) due to a much flatter temperature gradient that decreases the size of the nucleation free zone to a very low value (during traditional casting practices the alloy's Q value is a dominant factor in controlling grain size). This implies that nucleation is now controlled by the grain number density although a small effect of alloy composition remains evident (and in some conditions composition still has a significant effect [1]). These observations indicate that the interpretation of the term x_{Sd} in the Interdependence model is different for UST conditions. x_{Sd} is set by the number of grains accumulating in the melt as they are ejected from under the sonotrode whereas for normal casting conditions x_{Sd} is related to the number density of particle substrates that are able to be activated as nucleants within a bulk melt.

9. Directions for Future Research

While the application of UST in the liquid state has already been shown to be industrially useful [34], the application of UST across the liquidus shows potential as a method of obtaining very fine grain sizes but is much more difficult to implement commercially. With this in mind, consideration of the findings and mechanisms described in this paper suggest opportunities for future research.

Firstly, this work has highlighted that settling and casting size have a significant effect on the degree of grain refinement achieved. Work is needed to quantify the rate of grain formation during UST and the effect of settling on the grain size across the as-cast macrostructure. Related to this is the effect of density differences between grains and liquid on grain size as this will affect the degree of settling or a tendency for grains to float once UST is terminated. Further development of the Interdependence model as applied to UST conditions is needed to improve the prediction of the amount of refinement that can be achieved for particular alloys under specific casting conditions. Computer simulation of micro- and macro-structure development needs to take into account the movement of grains with convection and settling. As it was found that maintaining the sonotrode temperature below the liquidus temperature produced non-dendritic grains while, in the case of Zn, sonotrode heating occurred that reduced the number of non-dendritic grains, it would be worth exploring these effects on larger volumes or over longer times with a water-cooled sonotrode. In parallel, the effect of material properties such as thermal diffusion rate and heat capacity need to be understood to explain the formation of columnar grains enabling tailoring of alloy composition and casting conditions to prevent their formation. Our work has not been focused on the actual mechanisms of nucleation under the sonotrode. Real-time synchrotron studies have focused on cavitation. If this approach could detect nucleation events and the initial growth of grains (e.g., dendritic or non-dendritic) in real time it would clarify which of the proposed mechanisms in the literature are dominant.

Author Contributions: Conceptualisation of the paper D.S., N.B., and G.W.; Resources G.W. and M.D.; Methodology D.S., N.B., and G.W.; Formal analysis N.B.; Visualization, N.B.; Data curation N.B., G.W. and D.S.; Writing—Original draft preparation, N.B.; Writing—Review and Editing, N.B., D.S., M.D. and G.W.; Supervision, M.D. and G.W.; Project Administration, M.D.; Funding Acquisition, M.D. and G.W.

Funding: This research was funded by Australian Research Council (ARC) Research Hub for Advanced Manufacturing of Medical Devices, IH150100024, the ARC Discovery grant, DP140100702 and ARC linkage project, LP150100950. This paper includes research that was supported by DMTC Limited (Australia). The authors have prepared this paper in accordance with the intellectual property rights granted to partners from the original DMTC project.

Acknowledgments: The authors thank the valuable contributions from our collaborators Paul Croaker, Hajo Dieringa, Mark A. Easton, Dmitry G. Eskin, Damian McGuckin, Kurt Mills, Arvind Prasad, Ma Qian, Dong Qiu, Carmelo J. Todaro, Qiang Wang, Shiyang Liu, Albin Joseph and Andy Ngo at UQ, RMIT University and Helmholtz-Zentrum Geesthacht (HZG), who have assisted in undertaking the experiments and co-authoring our publications.

Conflicts of Interest: The authors declare no conflict of interest and the funders had no role in the data collection, analysis, writing and decision to publish the manuscript.

References

1. Ramirez, A.; Qian, M.; Davis, B.; Wilks, T.; StJohn, D.H. Potency of high-intensity ultrasonic treatment for grain refinement of magnesium alloys. *Scr. Mater.* **2008**, *59*, 19–22. [CrossRef]
2. Atamanenko, T.V.; Eskin, D.G.; Zhang, L.; Katgerman, L. Criteria of Grain Refinement Induced by Ultrasonic Melt Treatment of Aluminum Alloys Containing Zr and Ti. *Metall. Mater. Trans. A* **2010**, *41a*, 2056–2066. [CrossRef]
3. Qian, M.; Ramirez, A.; Das, A.; StJohn, D.H. The effect of solute on ultrasonic grain refinement of magnesium alloys. *J. Cryst. Growth* **2010**, *312*, 2267–2272. [CrossRef]
4. Qian, M.; Das, A. Grain refinement of magnesium alloys by zirconium: Formation of equiaxed grains. *Scr. Mater.* **2006**, *54*, 881–886. [CrossRef]
5. Qian, M.; Zheng, L.; Graham, D.; Frost, M.T.; StJohn, D.H. Settling of undissolved zirconium particles in pure magnesium melts. *J. Light Met.* **2001**, *1*, 157–165. [CrossRef]
6. Nagasivamuni, B.; Wang, G.; StJohn, D.H.; Dargusch, M.S. Effect of ultrasonic treatment on the alloying and grain refinement efficiency of a Mg-Zr master alloy added to magnesium at hypo- and hyper-peritectic compositions. *J. Cryst. Growth* **2019**, *512*, 20–32. [CrossRef]
7. Guan, R.G.; Tie, D. A Review on Grain Refinement of Aluminum Alloys: Progresses, Challenges and Prospects. *Acta Metall. Sin. (Engl. Lett.)* **2017**, *30*, 409–432. [CrossRef]
8. StJohn, D.H.; Qian, M.; Easton, M.A.; Cao, P.; Hildebrand, Z. Grain refinement of magnesium alloys. *Metall. Mater. Trans. A* **2005**, *36a*, 1669–1679. [CrossRef]
9. McCartney, D.G. Grain refining of aluminium and its alloys using inoculants. *Int. Mater. Rev.* **1989**, *34*, 247–260. [CrossRef]
10. StJohn, D.H.; Cao, P.; Qian, M.; Easton, M.A. A Brief History of the Development of Grain Refinement Technology for Cast Magnesium Alloys. In *Magnesium Technology 2013*; Hort, N., Mathaudhu, S.N., Neelameggham, N.R., Alderman, M., Eds.; Springer: Cham, Switzerland, 2013; pp. 3–8.
11. Easton, M.A.; Qian, M.; Prasad, A.; StJohn, D.H. Recent advances in grain refinement of light metals and alloys. *Curr. Opin. Solid State Mater. Sci.* **2016**, *20*, 13–24. [CrossRef]
12. Robson, J.D. Critical Assessment 9: Wrought magnesium alloys. *J. Mater. Sci. Technol.* **2015**, *31*, 257–264. [CrossRef]
13. StJohn, D.H.; Easton, M.A.; Qian, M.; Taylor, J.A. Grain Refinement of Magnesium Alloys: A Review of Recent Research, Theoretical Developments, and Their Application. *Metall. Mater. Trans. A* **2013**, *44a*, 2935–2949. [CrossRef]
14. Easton, M.; StJohn, D. Grain refinement of aluminum alloys: Part I. the nucleant and solute paradigms—a review of the literature. *Metall. Mater. Trans. A* **1999**, *30*, 1613–1623. [CrossRef]
15. StJohn, D.H.; Qian, M.; Easton, M.A.; Cao, P. The Interdependence Theory: The relationship between grain formation and nucleant selection. *Acta Mater.* **2011**, *59*, 4907–4921. [CrossRef]
16. Greer, A.L.; Bunn, A.M.; Tronche, A.; Evans, P.V.; Bristow, D.J. Modelling of inoculation of metallic melts: Application to grain refinement of aluminium by Al-Ti-B. *Acta Mater.* **2000**, *48*, 2823–2835. [CrossRef]
17. Bunn, A.M.; Schumacher, P.; Kearns, M.A.; Boothroyd, C.B.; Greer, A.L. Grain refinement by Al-Ti-B alloys in aluminium melts: A study of the mechanisms of poisoning by zirconium. *J. Mater. Sci. Technol.* **1999**, *15*, 1115–1123. [CrossRef]
18. Qiu, D.; Taylor, J.A.; Zhang, M.X.; Kelly, P.M. A mechanism for the poisoning effect of silicon on the grain refinement of Al-Si alloys. *Acta Mater.* **2007**, *55*, 1447–1456. [CrossRef]
19. Lee, Y.C.; Dahle, A.K.; StJohn, D.H.; Hutt, J.E.C. The effect of grain refinement and silicon content on grain formation in hypoeutectic Al-Si alloys. *Mater. Sci. Eng. A* **1999**, *259*, 43–52. [CrossRef]
20. Wang, G.; Dargusch, M.; Easton, M.; StJohn, D. Chapter 9—Treatment by External Fields. In *Fundamentals of Aluminium Metallurgy*; Lumley, R.N., Ed.; Woodhead Publishing: Sawston, UK, 2018; pp. 279–332.
21. Hiedemann, E.A. Metallurgical Effects of Ultrasonic Waves. *J. Acoust. Soc. Am.* **1954**, *26*, 831–842. [CrossRef]
22. Eskin, G.I. Cavitation mechanism of ultrasonic melt degassing. *Ultrason. Sonochem.* **1995**, *2*, S137–S141. [CrossRef]
23. Eskin, G.I. Broad prospects for commercial application of the ultrasonic (cavitation) melt treatment of light alloys. *Ultrason. Sonochem.* **2001**, *8*, 319–325. [CrossRef]

24. Todaro, C.J.; Easton, M.A.; Qiu, D.; Wang, G.; StJohn, D.H.; Qian, M. The Effect of Ultrasonic Melt Treatment on Macro-Segregation and Peritectic Transformation in an Al-19Si-4Fe Alloy. *Metall. Mater. Trans. A* **2017**, *48*, 5579–5590. [CrossRef]
25. Todaro, C.J.; Easton, M.A.; Qiu, D.; Wang, G.; StJohn, D.H.; Qian, M. Effect of ultrasonic melt treatment on intermetallic phase formation in a manganese-modified Al-17Si-2Fe alloy. *J. Mater. Process. Technol.* **2019**, *271*, 346–356. [CrossRef]
26. Eskin, D.G. Ultrasonic Melt Processing: Opportunities and Misconceptions. *Mater. Sci. Forum* **2014**, *794–796*, 101–106. [CrossRef]
27. Eskin, D.G. Ultrasonic processing of molten and solidifying aluminium alloys: Overview and outlook. *J. Mater. Sci. Technol.* **2017**, *33*, 636–645. [CrossRef]
28. Abramov, O.V. Action of high intensity ultrasound on solidifying metal. *Ultrasonics* **1987**, *25*, 73–82. [CrossRef]
29. Wang, F.; Eskin, D.; Mi, J.W.; Connolley, T.; Lindsay, J.; Mounib, M. A refining mechanism of primary Al3Ti intermetallic particles by ultrasonic treatment in the liquid state. *Acta Mater.* **2016**, *116*, 354–363. [CrossRef]
30. Wang, F.; Eskin, D.; Mi, J.W.; Wang, C.N.; Koe, B.; King, A.; Reinhard, C.; Connolley, T. A synchrotron X-radiography study of the fragmentation and refinement of primary intermetallic particles in an Al-35Cu alloy induced by ultrasonic melt processing. *Acta Mater.* **2017**, *141*, 142–153. [CrossRef]
31. Wang, F.; Tzanakis, I.; Eskin, D.; Mi, J.W.; Connolley, T. In situ observation of ultrasonic cavitation-induced fragmentation of the primary crystals formed in Al alloys. *Ultrason. Sonochem.* **2017**, *39*, 66–76. [CrossRef]
32. Wang, B.; Tan, D.Y.; Lee, T.L.; Khong, J.C.; Wang, F.; Eskin, D.; Connolley, T.; Fezzaa, K.; Mi, J.W. Ultrafast synchrotron X-ray imaging studies of microstructure fragmentation in solidification under ultrasound. *Acta Mater.* **2018**, *144*, 505–515. [CrossRef]
33. Wang, S.; Kang, J.; Zhang, X.; Guo, Z. Dendrites fragmentation induced by oscillating cavitation bubbles in ultrasound field. *Ultrasonics* **2018**, *83*, 26–32. [CrossRef] [PubMed]
34. Eskin, D.G.; Tzanakis, I.; Wang, F.; Lebon, G.S.B.; Subroto, T.; Pericleous, K.; Mi, J. Fundamental studies of ultrasonic melt processing. *Ultrason. Sonochem.* **2019**, *52*, 455–467. [CrossRef] [PubMed]
35. Shu, D.; Sun, B.; Mi, J.; Grant, P.S. A High-Speed Imaging and Modeling Study of Dendrite Fragmentation Caused by Ultrasonic Cavitation. *Metall. Mater. Trans. A* **2012**, *43*, 3755–3766. [CrossRef]
36. Eskin, D.G.; Tzanakis, I. High-Frequency Vibration and Ultrasonic Processing. In *Solidification Processing of Metallic Alloys under External Fields*; Eskin, D.G., Mi, J., Eds.; Springer: Cham, Switzerland, 2018; pp. 153–193.
37. Wang, G.; Croaker, P.; Dargusch, M.; McGuckin, D.; StJohn, D. Evolution of the As-Cast Grain Microstructure of an Ultrasonically Treated Al–2Cu Alloy. *Adv. Eng. Mater.* **2018**, *20*, 1800521. [CrossRef]
38. Wang, G.; Croaker, P.; Dargusch, M.; McGuckin, D.; StJohn, D. Simulation of convective flow and thermal conditions during ultrasonic treatment of an Al-2Cu alloy. *Comput. Mater. Sci.* **2017**, *134*, 116–125. [CrossRef]
39. Abramov, V.; Abramov, O.; Bulgakov, V.; Sommer, F. Solidification of aluminium alloys under ultrasonic irradiation using water-cooled resonator. *Mater. Lett.* **1998**, *37*, 27–34. [CrossRef]
40. Qian, M.; Ramirez, A.; Das, A. Ultrasonic refinement of magnesium by cavitation: Clarifying the role of wall crystals. *J. Cryst. Growth* **2009**, *311*, 3708–3715. [CrossRef]
41. Tan, D.Y.; Lee, T.L.; Khong, J.C.; Connolley, T.; Fezzaa, K.; Mi, J.W. High-Speed Synchrotron X-ray Imaging Studies of the Ultrasound Shockwave and Enhanced Flow during Metal Solidification Processes. *Metall. Mater. Trans. A* **2015**, *46a*, 2851–2861. [CrossRef]
42. Wang, S.; Kang, J.; Guo, Z.; Lee, T.L.; Zhang, X.; Wang, Q.; Deng, C.; Mi, J. In situ high speed imaging study and modelling of the fatigue fragmentation of dendritic structures in ultrasonic fields. *Acta Mater.* **2019**, *165*, 388–397. [CrossRef]
43. Nagasivamuni, B.; Wang, G.; StJohn, D.H.; Dargusch, M.S. The effect of ultrasonic treatment on the mechanisms of grain formation in as-cast high purity zinc. *J. Cryst. Growth* **2018**, *495*, 20–28. [CrossRef]
44. Nagasivamuni, B.; Wang, G.; StJohn, D.H.; Dargusch, M.S. Mechanisms of Grain Formation During Ultrasonic Solidification of Commercial Purity Magnesium. In *Light Metals 2019*; Chesonis, C., Ed.; Springer: Cham, Switzerland, 2019; pp. 1579–1586.
45. Wang, G.; Qiang Wang, E.; Prasad, A.; Dargusch, M.; StJohn, D.H. Grain Refinement Of Al-Si Hypoeutectic Alloys By Al_3Ti_1B Master Alloy And Ultrasonic Treatment. In *Shape Casting: 6th International Symposium*; Tiryakioğlu, M., Jolly, M., Byczynski, G., Eds.; Springer: Cham, Switzerland, 2016; pp. 143–150.

46. Wang, G.; Wang, Q.; Easton, M.A.; Dargusch, M.S.; Qian, M.; Eskin, D.G.; StJohn, D.H. Role of ultrasonic treatment, inoculation and solute in the grain refinement of commercial purity aluminium. *Sci. Rep.* **2017**, *7*, 9729. [CrossRef] [PubMed]
47. Wang, G.; Dargusch, M.S.; Eskin, D.G.; StJohn, D.H. Identifying the Stages during Ultrasonic Processing that Reduce the Grain Size of Aluminum with Added Al$_3$Ti$_1$B Master Alloy. *Adv. Eng. Mater.* **2017**, *19*. [CrossRef]
48. Prasad, A.; Yuan, L.; Lee, P.D.; StJohn, D.H. The Interdependence model of grain nucleation: A numerical analysis of the Nucleation-Free Zone. *Acta Mater.* **2013**, *61*, 5914–5927. [CrossRef]
49. Mills, K.; Wang, G.; StJohn, D.; Dargusch, M. Ultrasonic Processing of Aluminum–Magnesium Alloys. *Materials* **2018**, *11*, 1994. [CrossRef] [PubMed]
50. Dieringa, H.; Katsarou, L.; Buzolin, R.; Szakács, G.; Horstmann, M.; Wolff, M.; Mendis, C.; Vorozhtsov, S.; StJohn, D. Ultrasound Assisted Casting of an AM60 Based Metal Matrix Nanocomposite, Its Properties, and Recyclability. *Metals* **2017**, *7*, 388. [CrossRef]
51. Joseph, A. Solidification and Grain Refinement of Al-Cu and Al-Zn Alloys Using Ultrasonics, Al$_3$Ti$_1$B Grain Refiner and Their Combination. Bachelor's Thesis, The University of Queensland, Brisbane, Australia, 2017.
52. Wang, G.; Dargusch, M.S.; Qian, M.; Eskin, D.G.; StJohn, D.H. The role of ultrasonic treatment in refining the as-cast grain structure during the solidification of an Al-2Cu alloy. *J. Cryst. Growth* **2014**, *408*, 119–124. [CrossRef]
53. Wang, E.Q.; Wang, G.; Dargusch, M.S.; Qian, M.; Eskin, D.G.; StJohn, D.H. Grain Refinement of an Al-2 wt% Cu Alloy by Al$_3$Ti$_1$B Master Alloy and Ultrasonic Treatment. *IOP Conf. Ser. Mater. Sci. Eng.* **2016**, *117*, 012050. [CrossRef]
54. Mitome, H. The mechanism of generation of acoustic streaming. *Electron. Commun. Jpn. (Part III: Fundam. Electron. Sci.)* **1998**, *81*, 1–8. [CrossRef]
55. Hellawell, A.; Liu, S.; Lu, S.Z. Dendrite fragmentation and the effects of fluid flow in castings. *JOM* **1997**, *49*, 18–20. [CrossRef]
56. Campanella, T.; Charbon, C.; Rappaz, M. Grain refinement induced by electromagnetic stirring: A dendrite fragmentation criterion. *Metall. Mater. Trans. A* **2004**, *35*, 3201–3210. [CrossRef]
57. Tzanakis, I.; Lebon, G.S.B.; Eskin, D.G.; Pericleous, K.A. Characterizing the cavitation development and acoustic spectrum in various liquids. *Ultrason. Sonochem.* **2017**, *34*, 651–662. [CrossRef] [PubMed]
58. Nastac, L. Mathematical Modeling of the Solidification Structure Evolution in the Presence of Ultrasonic Stirring. *Metall. Mater. Trans. B* **2011**, *42*, 1297–1305. [CrossRef]
59. Shao, Z.; Le, Q.; Zhang, Z.; Cui, J. A new method of semi-continuous casting of AZ80 Mg alloy billets by a combination of electromagnetic and ultrasonic fields. *Mater. Des.* **2011**, *32*, 4216–4224. [CrossRef]
60. Lebon, G.S.B.; Tzanakis, I.; Djambazov, G.; Pericleous, K.; Eskin, D.G. Numerical modelling of ultrasonic waves in a bubbly Newtonian liquid using a high-order acoustic cavitation model. *Ultrason. Sonochem.* **2017**, *37*, 660–668. [CrossRef] [PubMed]
61. Fang, Y.; Yamamoto, T.; Komarov, S. Cavitation and acoustic streaming generated by different sonotrode tips. *Ultrason. Sonochem.* **2018**, *48*, 79–87. [CrossRef] [PubMed]
62. Lebon, G.S.B.; Tzanakis, I.; Pericleous, K.; Eskin, D.; Grant, P.S. Ultrasonic liquid metal processing: The essential role of cavitation bubbles in controlling acoustic streaming. *Ultrason. Sonochem.* **2019**, *55*, 243–255. [CrossRef] [PubMed]
63. Zhang, L.; Eskin, D.G.; Katgerman, L. Influence of ultrasonic melt treatment on the formation of primary intermetallics and related grain refinement in aluminum alloys. *J. Mater. Sci.* **2011**, *46*, 5252–5259. [CrossRef]
64. Sreekumar, V.M.; Eskin, D.G. A New Al-Zr-Ti Master Alloy for Ultrasonic Grain Refinement of Wrought and Foundry Aluminum Alloys. *JOM* **2016**, *68*, 3088–3093. [CrossRef]
65. Easton, M.; StJohn, D. An analysis of the relationship between grain size, solute content, and the potency and number density of nucleant particles. *Metall. Mater. Trans. A* **2005**, *36*, 1911–1920. [CrossRef]
66. Sun, M.; Easton, M.A.; StJohn, D.H.; Wu, G.H.; Abbott, T.B.; Ding, W.J. Grain Refinement of Magnesium Alloys by Mg-Zr Master Alloys: The Role of Alloy Chemistry and Zr Particle Number Density. *Adv. Eng. Mater.* **2013**, *15*, 373–378. [CrossRef]
67. Han, Y.F.; Shu, D.; Wang, J.; Sun, B.O. Microstructure and grain refining performance of Al-5Ti-1B master alloy prepared under high-intensity ultrasound. *Mater. Sci. Eng., A* **2006**, *430*, 326–331. [CrossRef]
68. Wang, F.; Eskin, D.; Connolley, T.; Mi, J. Effect of ultrasonic melt treatment on the refinement of primary Al3Ti intermetallic in an Al-0.4Ti alloy. *J. Cryst. Growth* **2016**, *435*, 24–30. [CrossRef]

69. Atamanenko, T.V.; Eskin, D.G.; Sluiter, M.; Katgerman, L. On the mechanism of grain refinement in Al-Zr-Ti alloys. *J. Alloys Compd.* **2011**, *509*, 57–60. [CrossRef]
70. Kotadia, H.R.; Das, A. Modification of solidification microstructure in hypo- and hyper-eutectic Al-Si alloys under high-intensity ultrasonic irradiation. *J. Alloys Compd.* **2015**, *620*, 1–4. [CrossRef]
71. Jung, J.-G.; Lee, J.-M.; Cho, Y.-H.; Yoon, W.-H. Combined effects of ultrasonic melt treatment, Si addition and solution treatment on the microstructure and tensile properties of multicomponent AlSi alloys. *J. Alloys Compd.* **2017**, *693*, 201–210. [CrossRef]
72. Kim, S.-B.; Cho, Y.-H.; Jung, J.-G.; Yoon, W.-H.; Lee, Y.-K.; Lee, J.-M. Microstructure-Strengthening Interrelationship of an Ultrasonically Treated Hypereutectic Al-Si (A390) Alloy. *Met. Mater. Int.* **2018**, *24*, 1376–1385. [CrossRef]
73. Harini, R.S.; Nampoothiri, J.; Nagasivamuni, B.; Raj, B.; Ravi, K.R. Ultrasonic assisted grain refinement of Al-Mg alloy using in-situ MgAl$_2$O$_4$ particles. *Mater. Lett.* **2015**, *145*, 328–331. [CrossRef]
74. Kotadia, H.R.; Qian, M.; Eskin, D.G.; Das, A. On the microstructural refinement in commercial purity Al and Al-10 wt% Cu alloy under ultrasonication during solidification. *Mater. Des.* **2017**, *132*, 266–274. [CrossRef]
75. Sreekumar, V.M.; Babu, N.H.; Eskin, D.G. Potential of an Al-Ti-MgAl$_2$O$_4$ Master Alloy and Ultrasonic Cavitation in the Grain Refinement of a Cast Aluminum Alloy. *Metall. Mater. Trans. B* **2017**, *48*, 208–219. [CrossRef]
76. Wang, G.; Wang, Q.; Balasubramani, N.; Qian, M.; Eskin, D.G.; Dargusch, M.S.; StJohn, D.H. The Role of Ultrasonically Induced Acoustic Streaming in Developing Fine Equiaxed Grains During the Solidification of an Al-2Pct Cu Alloy. *Metall. Mater. Trans. A* **2019**, 1–11. [CrossRef]
77. Lebon, G.S.B.; Salloum-Abou-Jaoude, G.; Eskin, D.; Tzanakis, I.; Pericleous, K.; Jarry, P. Numerical modelling of acoustic streaming during the ultrasonic melt treatment of direct-chill (DC) casting. *Ultrason. Sonochem.* **2019**, *54*, 171–182. [CrossRef] [PubMed]
78. Wang, S.; Guo, Z.P.; Zhang, X.P.; Zhang, A.; Kang, J.W. On the mechanism of dendritic fragmentation by ultrasound induced cavitation. *Ultrason. Sonochem.* **2019**, *51*, 160–165. [CrossRef] [PubMed]
79. Srivastava, N.; Chaudhari, G.P.; Qian, M. Grain refinement of binary Al-Si, Al-Cu and Al-Ni alloys by ultrasonication. *J. Mater. Process. Technol.* **2017**, *249*, 367–378. [CrossRef]
80. Feng, X.H.; Zhao, F.Z.; Jia, H.M.; Zhou, J.X.; Li, Y.D.; Li, W.R.; Yang, Y.S. Effect of temperature conditions on grain refinement of Mg-Al alloy under ultrasonic field. *Int. J. Cast Met. Res.* **2017**, *30*, 341–347. [CrossRef]
81. Nowak, M.; Bolzoni, L.; Hari Babu, N. Grain refinement of Al-Si alloys by Nb–B inoculation. Part I: Concept development and effect on binary alloys. *Mater. Des. (1980–2015)* **2015**, *66*, 366–375. [CrossRef]
82. StJohn, D.H.; Prasad, A.; Easton, M.A.; Qian, M. The Contribution of Constitutional Supercooling to Nucleation and Grain Formation. *Metall. Mater. Trans. A* **2015**, *46*, 4868–4885. [CrossRef]
83. Prasad, A.; Liotti, E.; McDonald, S.D.; Nogita, K.; Yasuda, H.; Grant, P.S.; StJohn, D.H. Real-time synchrotron x-ray observations of equiaxed solidification of aluminium alloys and implications for modelling. *IOP Conf. Ser. Mater. Sci. Eng.* **2015**, *84*, 012014. [CrossRef]
84. Guo, E.; Shuai, S.; Kazantsev, D.; Karagadde, S.; Phillion, A.B.; Jing, T.; Li, W.; Lee, P.D. The influence of nanoparticles on dendritic grain growth in Mg alloys. *Acta Mater.* **2018**, *152*, 127–137. [CrossRef]
85. Chen, L.-Y.; Xu, J.-Q.; Li, X.-C. Controlling Phase Growth During Solidification by Nanoparticles. *Mater. Res. Lett.* **2015**, *3*, 43–49. [CrossRef]

© 2019 by the authors. Licensee MDPI, Basel, Switzerland. This article is an open access article distributed under the terms and conditions of the Creative Commons Attribution (CC BY) license (http://creativecommons.org/licenses/by/4.0/).

Article

Contactless Ultrasonic Cavitation in Alloy Melts

Koulis Pericleous [1,*], Valdis Bojarevics [1], Georgi Djambazov [1], Agnieszka Dybalska [2], William D. Griffiths [2] and Catherine Tonry [1]

[1] Centre for Numerical Modelling and Process Analysis, University of Greenwich, London SE10 9LS, UK; v.bojarevics@gre.ac.uk (V.B.); g.djambazov@gre.ac.uk (G.D.); C.Tonry@gre.ac.uk (C.T.)
[2] School of Metallurgy and Materials, University of Birmingham, Birmingham B15 2TT, UK; A.Dybalska@bham.ac.uk (A.D.); W.D.Griffiths@bham.ac.uk (W.D.G.)
* Correspondence: k.pericleous@gre.ac.uk

Received: 13 September 2019; Accepted: 31 October 2019; Published: 3 November 2019

Abstract: A high frequency tuned electromagnetic induction coil is used to induce ultrasonic pressure waves leading to cavitation in alloy melts. This presents an alternative '*contactless*' approach to conventional immersed probe techniques. The method can potentially offer the same benefits of traditional ultrasonic treatment (UST) such as degassing, microstructure refinement and dispersion of particles, but avoids melt contamination due to probe erosion prevalent in immersed sonotrodes, and it can be used on higher temperature and reactive alloys. An added benefit is that the induction stirring produced by the coil, enables a larger melt treatment volume. Model simulations of the process are conducted using purpose-built software, coupling flow, heat transfer, sound and electromagnetic fields. Modelling results are compared against experiments carried out in a prototype installation. Results indicate strong melt stirring and evidence of cavitation accompanying acoustic resonance. Up to 63% of grain refinement was obtained in commercial purity (CP-Al) aluminium and a further 46% in CP-Al with added Al–5Ti–1B grain refiner.

Keywords: ultrasonic treatment; contactless sonotrode; induction processing; grain refinement

---

1. Introduction

Ultrasonic treatment (UST) of molten metals prior to solidification has been shown to improve their mechanical properties refining microstructure, degassing and dispersing strengthening particles [1]. The standard process involves the use of an immersed sonotrode probe vibrating at ultrasonic frequency (~20 kHz). Intense pressure waves are generated, which trigger dissolved gas cavitation that in turn assists nucleation [2], causes the break-up of intermetallics and evolving dendrites [3] and disperses clusters of particles. The process is, so far, mainly applied in low temperature melts (e.g., Al, Mg) but even then, there are problems preventing its widespread use by industry. The probe tip is dissolved at varying rates leading to melt contamination. The amount of liquid metal treated is concentrated in a small volume surrounding the probe, which necessitates additional mechanical stirring in order to spread the effect to a larger melt volume. Multiple sonotrodes often need to be used, making scalability complex and expensive [1].

To avoid these problems we devised the novel contactless electromagnetic (EM) UST process presented here. The new process eliminates the risk of melt contamination and the associated cost of frequent probe replacement, opening the potential benefits of UST to high temperature (e.g., Ni, Fe, Cu, ODS steel) or reactive (e.g., Ti, Zr) alloys. In contrast to the immersed sonotrode technique, where the kinetic energy of the vibrating horn is directly interacting with the liquid, the contactless EM device relies on acoustic resonance to reach pressure amplitudes leading to cavitation [4]. Furthermore, the Lorentz force, due to the induced current, leads to strong stirring, promoting multiple passages of the melt through active '*Blake threshold*' pressure zones. Numerical simulations [4] indicate that scaling up to larger volumes can be achieved by simply introducing a larger coil and adjusting the current or frequency to match the acoustic resonance characteristics of the melt volume and surrounding container. Studying earlier publications [5,6], the idea of applying static or AC magnetic fields for the contactless ultrasonic treatment of liquid metals is not new, but the present implementation remains unique both in concept and in the ease with which it can be implemented in industry.

The remainder of the paper introduces the device known as the 'Top Coil' contactless sonotrode, summarises the mathematical methods used to model its function, followed by the experimental procedure together with sample results and discussion. Concluding remarks and references follow.

2. The Contactless Sonotrode

The patented [7] 'Top Coil' sonotrode consists of a conical induction coil that can be lowered into the crucible containing the molten alloy, as shown schematically in Figure 1. The coil is water-cooled, with a current through it of sufficient magnitude to create a gap by EM repulsion between the liquid metal and the coil surface (typically ~1700 kA at ~9.5 kHz has been used for aluminium). A protective ceramic coating is employed as an additional safety feature to eliminate the risk of spark erosion.

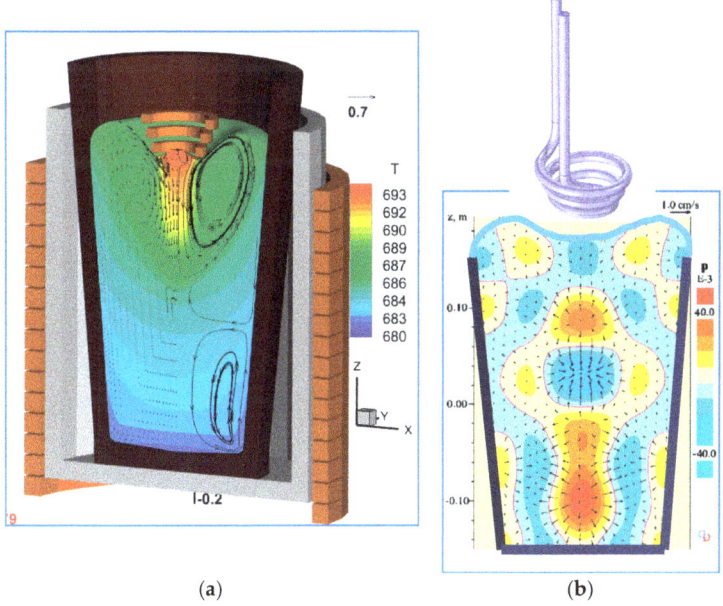

Figure 1. Simulations demonstrate the sonotrode concept [7]: (**a**) velocity and temperature <680–693 °C>, (**b**) instantaneous sound field < ± 40 kPa > with coil operating at 1000 A, 10 kHz.

Time-dependent simulations using a purpose-built spectral collocation code [8,9] coupling magnetic fields, turbulent flow and heat transfer in a dynamically varying fluid volume were used to design this system [4,7]. Details of the mathematical model used are given in Section 3 below. A typical simulation result in Figure 1, shows the dual effect of the Lorentz force arising from the interaction between the coil current and the opposing current induced in the melt: (i) In Figure 1a, the time-averaged component of the force repels the free surface and generates strong bulk stirring, (ii) the time-dependent component, acting at twice the supply frequency (see Equation (10)) vibrates the melt generating sound waves. The generation of sound waves is important in the process, since at the frequency needed for ultrasonic operation (~20 kHz), the applied EM force is concentrated in a thin 'skin layer' of fluid on the free surface (~7 mm in Al). Its effect can only be transmitted to the bulk through sound waves as pressure fluctuations, as shown in Figure 1b. In this case, the sound field was obtained by solving the compressible Euler momentum equations using a 4th order accurate finite difference scheme [10]. The amplitude of pressure fluctuations determines whether gas bubbles will emerge out of the solution, oscillate and under certain conditions cavitate. To reach the necessary pressure threshold for cavitation, the frequency of the coil needs to be tuned to produce resonance in the treatment vessel, accounting for the vessel geometry, free surface shape, temperature and crucible sound absorption characteristics. The use of resonance to enhance pressure amplitudes reduces the need for a very high current in the coil, as is found to be necessary in other proposed EM vibration techniques, for example in [5,6]. This latter fact makes the process energy efficient, especially where industrial scale operations are to be considered. The sound field simulations are needed to guide the frequency selection within the bounds of the power supply capacity.

In contrast to the immersed sonotrode technique where the cavitation energy is concentrated in a conical region surrounding the probe, in the proposed method, most cavitation activity is expected to lie in resonant nodes deep in the melt, were induced flow stirring will ensure that gas bubbles can have multiple passes through, improving cavitation efficiency.

3. Mathematical Basis

It can be seen from the process description that this is a multi-physics application encompassing a range of traditional engineering fields. Due to space limitations the essential features of the models used are given here in summary with more detailed mathematical formulations given in the accompanying references.

The set of equations representing fluid dynamics and heat transfer are solved using a spectral collocation scheme [8] on a dynamically varying solution grid [9] covering the liquid metal volume. The soundfield calculation domain includes the crucible and surrounding ambient region computed in the time domain, using a 4th order staggered variable scheme on a regular Cartesian grid [10]. Cavitation alters the speed of sound locally as the appearance of gas alters the medium compressibility. This aspect of the problem is handled using an extension of the Rayleigh–Plesset equations as suggested by Caflish [11] and implemented in [12,13].

3.1. Turbulent Fluid Flow and Heat Transfer

Characterised by the momentum and mass continuity equations, given by:

$$\partial_t \mathbf{v} + (\mathbf{v}.\nabla)\mathbf{v} = -\rho^{-1}\nabla p + \nabla \cdot (\nu_e(\nabla\mathbf{v} + \nabla\mathbf{v}^T)) + \rho^{-1}\mathbf{j} \times \mathbf{B} + \mathbf{g} \qquad (1)$$

$$\nabla \cdot \mathbf{v} = 0 \qquad (2)$$

where, $\mathbf{v}$ is the velocity vector, p the pressure, ρ the density, ν_e the effective viscosity (sum of laminar and turbulent contributions), $\mathbf{j}$ the current density, $\mathbf{B}$ the magnetic field density, and $\mathbf{g}$ the gravity constant. The $\mathbf{j} \times \mathbf{B}$ term in Equation (1) represents the volumetric Lorentz force acting on the fluid.

$$C_p(\partial_t T + \mathbf{v} \cdot \nabla T) = \nabla \cdot (C_p \alpha_e \nabla T) + \rho^{-1}|\mathbf{J}|_2/\sigma \qquad (3)$$

where, T is the temperature, C_p the specific heat, α_e the effective thermal diffusivity and σ the electrical resistivity of the liquid. The last term in Equation (3) represents the Joule heating generated by the induced current in the metal.

Turbulence is modelled by the k-ω Turbulence Model [14] (including magnetic field interaction):

$$\partial_t k + \mathbf{v} \cdot \nabla k = \nabla \cdot [(\nu + \sigma_k \nu_T)\nabla k] + G - \beta^* \omega k - \frac{2\alpha_m k}{\rho/(\sigma \mathbf{B}^2)} \qquad (4)$$

$$\partial_t \omega + \mathbf{v} \cdot \nabla \omega = \nabla \cdot [(\nu + \sigma_\omega \nu_T)\nabla \omega] + \alpha \frac{\omega}{k} G - \beta \omega^2 - \frac{\alpha_m \omega}{\rho/(\sigma \mathbf{B}^2)} \qquad (5)$$

where, k is the kinetic energy of turbulence and ω its rate of dissipation. Note, standard nomenclature and model constants are used as in Wilcox [14].

3.2. Magnetic Induction

The AC magnetic field of angular frequency w due to the coil, B, and the induced current, J, can be divided into real and imaginary components.

$$B = B_R \cos \omega t + B_I \sin \omega t \tag{6}$$

$$J = J_R \cos \omega t + J_I \sin \omega t \tag{7}$$

where,

$$J_R = \sigma \frac{\omega}{2} \delta (B_R + B_I); \ J_I = \sigma \frac{\omega}{2} \delta (-B_R + B_I) \tag{8}$$

The skin depth indicating current penetration into the fluid is given by

$$\delta = \sqrt{\left(\frac{2}{\mu \omega \sigma}\right)} \tag{9}$$

where the magnetic permeability

$$\mu (= \mu 0) = 4\pi \times 10 - 7 H/m$$

The Lorentz force is given by:

$$\begin{aligned} F &= J \times B = \overline{F} + \widetilde{F} \\ \overline{F} &= \tfrac{1}{2}(J_R B_R + J_I B_I) = \tfrac{1}{2\mu \delta} B_o^2 e^{-2\frac{x}{\delta}} \\ \widetilde{F} &= \tfrac{1}{2\mu \delta} B_o^2 e^{-2\frac{x}{\delta}} \sqrt{2} \cos(2\omega t - 2\tfrac{x}{\delta} + \tfrac{\pi}{4}) \end{aligned} \tag{10}$$

As shown in Equation (10), the Lorentz force F being the cross product of magnetic field and current can be divided into mean and time-dependent (sinusoidal) components. The mean value is responsible for bulk stirring, while the sinusoidal part is the source of vibration. It is important to note (a) that the vibration frequency is double that of the supply current, (b) that the force decays within the skin-depth distance δ from the liquid free surface, hence the importance of resonance for achieving the required pressure amplitude for cavitation in the bulk volume.

Once the charge has melted with the aid of the main furnace coil surrounding the crucible, the top coil is moved axially down towards the free surface, in order to increase the EM coupling. The model simulations in Figure 2 show the process at three different time steps as the coil gradually deflects the free surface, together with the associated flow field and temperature distribution. The main furnace coil surrounds the crucible and contributes to stirring leading to the toroidal vortex pair appearing in the first two images. The top coil contributes to the melt temperature due to Joule heating, therefore to maintain the temperature at the optimum value for cavitation (in the case of aluminium around 700 °C), at some point in the process the furnace coil is turned off. In Figure 2, the stirring pattern is seen to change when this happens, as in the third figure on the right, where a single vortex dominates leading to deep recirculation in the crucible. Table 1 contains a working set of material properties used in the simulations.

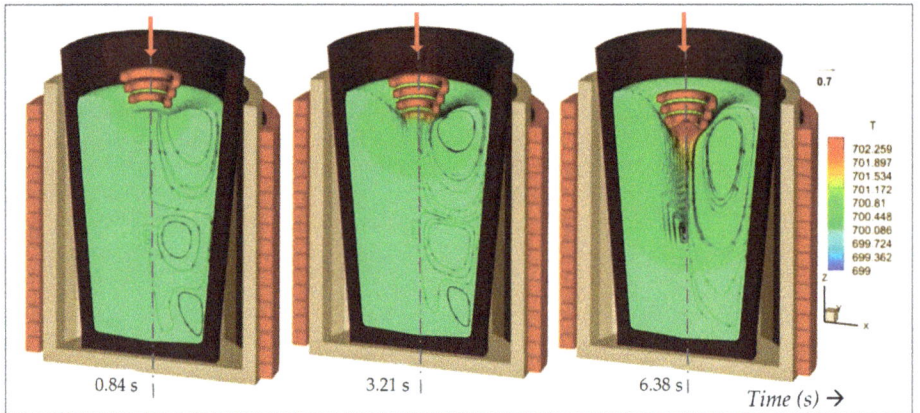

Figure 2. Predicted induced stirring and Joule heating in a typical aluminium crucible interacting with a descending conical coil. The main furnace coil operating at 2.4 kHz is switched of when the coil is in the lowest position, to maintain a maximum temperature of ~700 °C. Indicated temperature contour range < 699–702 °C. Maximum induced velocity is ~0.7 m/s.

Table 1. Material properties of aluminium.

Material Property	Aluminium (700 °C)
Sound Speed c (m s^{-1})	4600
Density ρ (kg m^{-3})	2350
Dynamic Viscosity μ (mPa s)	1.3
Surface Tension γ (N m^{-1})	0.87
Thermal Conductivity λ (Wm^{-1}K^{-1})	92
Electrical Conductivity σ (Sm^{-1})	3.8×10^7
Specific Heat C_p (kJ kg^{-1} K^{-1})	1.18

3.3. Soundfield Computation

The mean Lorentz force component $\overline{F}$ is responsible for bulk stirring. The time-dependent component $\widetilde{F}$, is the source of sound waves, computed by solving the Euler form of the momentum equations, generating a perturbation velocity field $\widetilde{v}$:

$$\frac{\partial p}{\partial t} + \rho c^2 \frac{\partial \widetilde{v}}{\partial x} = S;\ \rho \frac{\partial \widetilde{v}}{\partial t} + \frac{\partial p}{\partial x} = \widetilde{F} \tag{11}$$

A staggered scheme (in space and time) is used to solve Equation (11), with details are given in Djambazov et al. [10]. The source S represents pressure contributions due to cavitating bubbles [13]. The solution domain for sound extends beyond the melt to include the crucible and surrounding structures, thereby taking into account transmission and reflection of sound through the crucible walls. This means the acoustic impedance of all materials present needs to be considered. Constant pressure is assumed at the liquid free surface and a sound-hard boundary (zero flux) is applied at the coil surface. Details of the approach are given in [15].

A characteristic of the new process is the appearance of pressure nodes/antinodes deep inside the melt volume marking likely cavitation regions (e.g., see Figure 1b). This contrasts with the immersed sonotrode case, where cavitation activity is restricted to an area surrounding the vibrating probe, leading to shielding effects that limit process efficiency.

4. Experimental Methods

Grain refinement experiments have been carried out using a cylindrical crucible investigating the top coil performance for CP-Al with and without added Al–5Ti–1B grain refiner. Numerical simulations were used in each case, to compute the optimum frequency for resonance, taking into account the melt volume, crucible geometry and acoustic properties of all materials present. Since the efficiency of the UST process depends on the extent of gas cavitation, a means of detecting cavitation activity in opaque liquids is necessary. In parallel research using immersed sonotrodes, we employed a specially commissioned cavitometer that operates through a long tungsten rod (the probe), providing thermal protection to the piezo sensing elements placed well outside the hot area, and with a bandwidth capable of capturing broadband acoustic emissions associated with cavitation activity [16]. Due to inductive pickup, the cavitometer could not be used with the top coil, relying instead on an externally mounted digital high frequency microphone, Ultramic®200K, to record sound emitted from the crucible and thereby detect cavitation activity. The experimental setup is shown in Figure 3. For reference, Table 2, gives the composition of alloys tested in various experiments.

Figure 3. The 'top-coil' arrangement, showing the location of the high frequency microphone relative to the crucible.

Table 2. Composition of alloys tested.

Alloy	Si	Mg	Ti	Cu	Fe	Be	Mn	Zn	Balance Al
A357	6.5–7.5	0.55–0.6	0.1–0.2	0.0–0.2	0.1	0.002	0.1	0.0–0.1	90.8–93.0
CP-Al	0.3	0.03	0.0	0.03	0.4	0.0	0.03	0.07	99.5

4.1. Flowfield Validation

Figure 4a shows the computed velocity and temperature field in a small conical 125 mm crucible containing A357 alloy used in experiments to follow the tracks of radioactive particles, using the PEPT technique [17,18] and Figure 4b shows, in a typical experimental result, the aluminium surface with the coil immersed in it. As predicted by the model, the free surface of the melt is depressed by the Lorentz force, preventing contact with the coil. Along the axis of the coil, the EM force vanishes, leading to the conical elevation of the surface, shown in both simulation and experiment. Strong radial motion was observed just below the thin layer of oxide (Figure 4b) in all experiments. This gave qualitative support to model flow predictions, of a dominant toroidal vortex pushing the liquid down close to the axis, and then returning near the periphery.

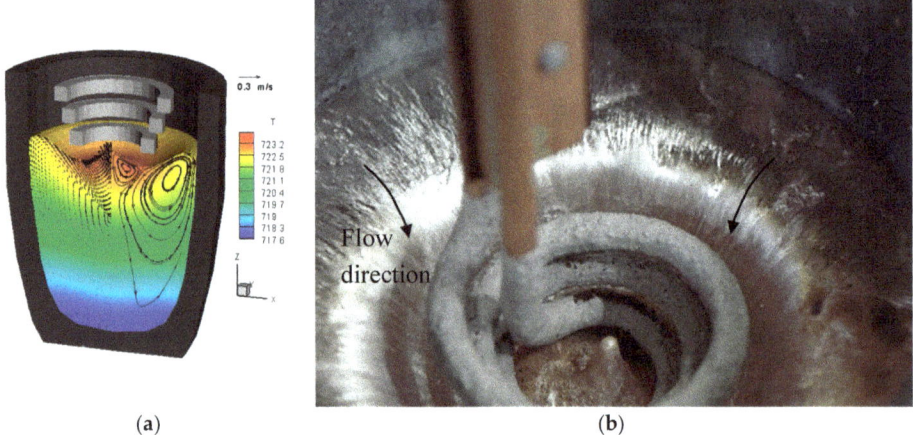

Figure 4. (**a**) Computed flow and heat transfer in a small 125 mm experimental crucible, containing A357 aluminium alloy used for particle tracking studies, (**b**) experiment showing the free surface of the melt; radial striations on the oxide layer indicate the flow direction and the conical elevation coinciding with the coil axis. Temperature contour range <717–723 °C>, maximum velocity 0.3 m/s.

The extent of stirring and therefore the ability of the top coil to disperse particles in the melt is clearly demonstrated in Figure 5. The numerical result shows 100 µm particle tracks obtained using Lagrangian tracking, accounting for the effects of turbulence and electromagnetic Kolin-Leenov forces [19]. In the simulation, particles seeded near the geometrical centre of the crucible are rapidly dispersed throughout the melt. The experimental result shows a similar dispersion pattern, obtained by tracking 200 µm radioactive particles using the Positron Emission Particle Tracking (PEPT) technique. From the processing point of view, this rigorous mixing is significant, since following initial cavitation, bubble or oxide fragments have the opportunity to re-enter the cavitation zone multiple times improving volumetric nucleation efficiency.

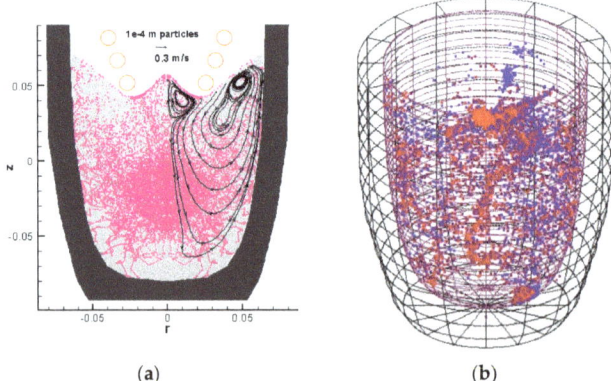

Figure 5. PEPT experiment, 125 mm crucible: (**a**) Simulation result showing dispersion of 100 µm particles due to induction stirring, (**b**) experimental 200 µm radioactive particle traces obtained using the PEPT technique.

4.2. Grain Refinement

To evaluate the grain refining potential of the process, experiments were conducted in a cylindrical clay-graphite crucible with internal and external diameters of 135 and 170 mm respectively, and depth 280 mm. For each experiment the crucible was filled with about 8.5 kg metal, either commercial purity aluminium (CP-Al), or CP-Al with the addition of 0.2 wt.% Al–5Ti–1B grain refiner (100 ppm Ti, 20 ppm B). The top coil was positioned centrally above the liquid metal surface and during processing the ambient ultrasonic noise emitted around the crucible was recorded using the Ultramic®200K digital ultrasonic microphone. Recorded sound was observed in the form of a FFT (Fast Fourier transform) sound spectrum extracted in real-time during experiments using MatLab®software (R2014a). As mentioned earlier, detection of cavitation in an opaque medium is a non-trivial problem; one indicator of cavitation was the presence of broadband noise emitted by collapsing bubbles [20,21]. This was seen (Figure 6a) in the form of light-coloured vertical lines on spectrograms recorded over a period of 1–2 min. The lines appear normal to the continuous horizontal lines denoting the top-coil frequency signal, observed at around 20 kHz, and the induction furnace signal, observed at around 5 kHz. The number and density of vertical lines was considered to be a good indication of cavitation activity [22].

The conditions generating the noise, (coil frequency and melt temperature), were then maintained for a further 5 min to produce samples for grain structure analysis. The intensity of cavitation during the process should be reflected in the grain structure of the samples, which were taken using the KBI ring test [23]. In this test, liquid metal is poured into a steel ring with an outside diameter of 75 mm, inside diameter 50 mm and height 25 mm, placed on an insulating silica brick. The cast sample is then subject to three simultaneous modes of cooling: through the air, the steel mould, and the silica brick. As the tuned resonant frequencies were shown by the simulations to be dependent on melt volume [22], the KBI ring test was the most useful for the contactless sonotrode experiments, as it only requires small samples, of about 50 g of Al. For grain size characterization, the base of the cylindrical samples was removed to about 3 mm above the bottom and ground, polished and etched with either Poultons' or Kellers' solution. The average grain size was then determined using the mean linear intercept method.

Figures 6 and 7 show the recorded spectrograms and post mortem grain structures of samples obtained following ultrasound treatment.

Figure 6b,c show the grain refinement achieved with CP-Al at 700 °C in a 150 mm diameter cylindrical crucible with 1700 A, 9.35 kHz current through the coil, corresponding to the spectrogram in Figure 6a. It was found that one of the factors that had to be controlled during ultrasound processing was the melt temperature, which must be kept low to promote cavitation. In these experiments it was maintained at 40 °C above the melting point, which for pure Al was 700 °C, the minimum value at which it was possible to pour the liquid metal [24] in a casting. Since the cavitation process starts with the formation of bubbles, the solubility of hydrogen gas in the liquid Al is an important factor in the process. Solubility decreases with temperature, so at lower temperatures, the existence of stable bubbles is more probable [1,25]. In the case of CP-Al (Figure 6), the grain size reduction was about 63%, (a reduction from 256 ± 12 to 95 ± 1 µm). This level of performance is consistent with previous findings where 70% grain size reduction was obtained [22], measured, in that case, by using the Aluminium Association Test Procedure-1 (TP1).

There are several reasons given in the literature for the observed reduction in grain size due to cavitation. Cavitation is believed to induce heterogeneous nucleation by (i) forced wetting of non-wetted particles present in the melt, resulting in an increased number of nucleation sites [26,27], (ii) local undercooling due to pressure changes when bubbles collapse [28] or (iii) undercooling of the melt at the bubble surface when the bubble rapidly expands [29]. In the case of CP-Al, the number of non-wetted particles should be smaller than in the case of alloys with grain refiner addition, a fact that makes grain refinement more difficult.

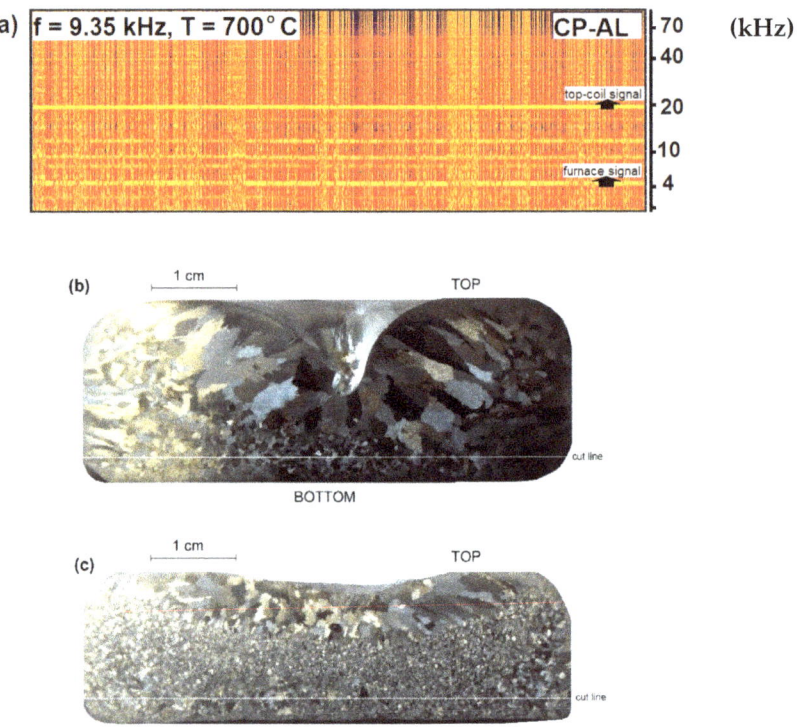

Figure 6. Grain refinement observed in CP-Al. (**a**) Recorded broadband noise during processing, (**b**) unprocessed sample, (**c**) sample processed by the contactless sonotrode at a frequency of 9.35 kHz at 700 °C. The cut line indicates the plane used for grain size measurements.

The addition of grain refiner increased the number of active nuclei and therefore all three mechanisms of cavitation-induced heterogeneous nucleation can take place. Figure 7 shows the grain size reduction achieved for CP-Al with a grain refiner. Grain sizes decreased from 223 ± 5 to 121 ± 2 µm.

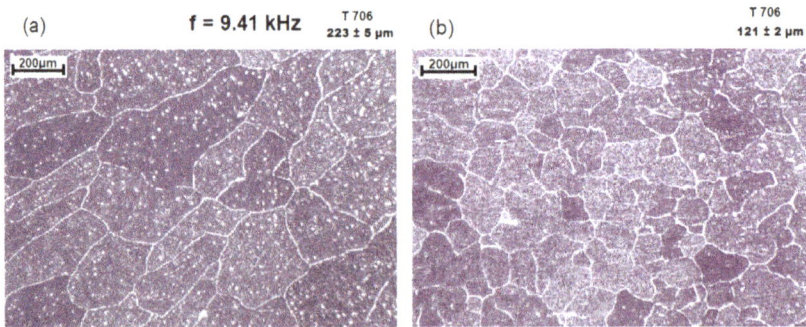

Figure 7. Grain size reduction following processing of sample containing commercial grain refiner: (**a**) the sample with Al–5Ti–1B addition, (**b**) alloy additionally processed by the contactless sonotrode with frequency 9.41 kHz for about 5 mins (Both samples were cast at 706 °C).

The experiments used a 0.2% Al–5Ti–1B ternary master alloy, which is commonly adopted as a grain refiner for most aluminium alloys [30,31]. Using the same alloy section as in Figure 6, grain sizes of the base alloy are shown in Figure 7a and the reduction caused by ultrasound shown in Figure 7b.

4.3. Correlation Between Grain Refinement and Frequency Spectrum

The basic concept behind the contactless sonotrode relies on the initiation of gas cavitation activity using acoustic resonance as the main driver for grain refinement. Numerical simulations provide an indication of the likely resonant modes given the sound properties and geometry of the alloy and crucible materials [15,32]. However, since these properties can vary unpredictably (especially so in ceramic crucibles), the experiment traverses the space about the indicated central frequency value using the spectrogram as an indicator of the most potent value, judged by the frequency of broadband noise bursts. Examining the results obtained in the larger 140 mm diameter crucible with internal depth 300 mm, we see for example, with reference to Figure 8, that spectrogram (a) (coil frequency 9.32 kHz) shows no cavitation activity, whilst spectrogram (b) (coil frequency 9.42 kHz) shows a dense pattern of cavitation bursts. Also evident in all spectrograms is that the cavitation activity as shown by the vertical lines is intermittent. This may be due to local changes in sound velocity in the melt as clouds of bubbles appear and disappear, which would disrupt the resonant conditions.

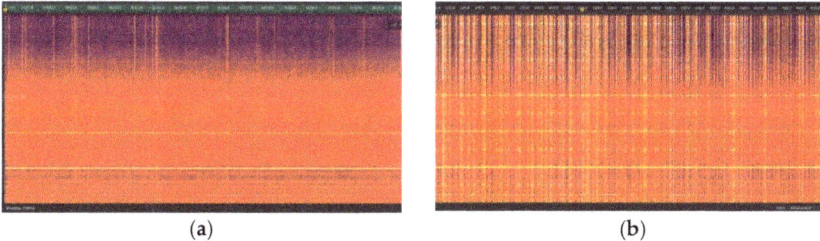

(a)   (b)

Figure 8. Contrasting nature of cavitation in two cases with very similar generator frequency, in (**a**) 9.32 kHz, in (**b**) 9.42 kHz obtained in a crucible with internal diameter 140 mm (remembering the vibration frequency in the coil will be doubled in the melt (10)).

It is then interesting to examine the sound wave resonant nodes that are most likely to excite cavitation. Figure 9 scans the range between 9.32 and 9.56 kHz applied to the 140 mm diameter cylindrical crucible containing CP-Al. In Figure 9a a typical FFT for the fairly active 9.43 kHz experiment identifies strong sound peaks at the driving frequency fo and its 3rd and 5th harmonics, an indicator of axial (up and down) wave reflections. The radial waves identified by the even harmonics are much weaker. Figure 9b Shows the spectrogram for the 9.43 kHz case, identifying the various peaks as horizontal lines (i.e., persisting in time). Finally, in Figure 9c the peak amplitude for the various harmonics was plotted against the driving frequency. The cavitation region coincides with the bulge in the 3rd and 5th harmonics amplitude, between 18,900 Hz and 19,100 Hz.

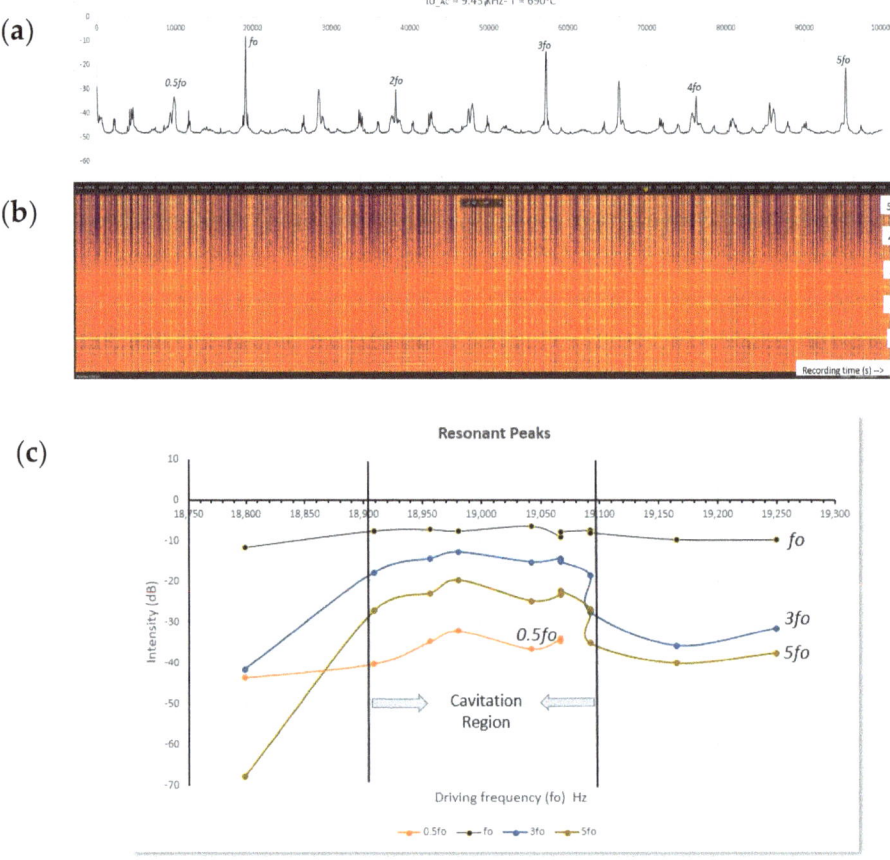

Figure 9. (a) FFT spectrum showing resonant peaks for 9.43 kHz (nominal) coil current, (b) corresponding spectrogram identifying horizontal lines as functions of driving frequency, (c) resonant peak intensity variation versus driving frequency (18,800–19,250 Hz), showing the cavitation region.

5. Concluding Remarks

This paper shows a contactless electromagnetic processing technique that can generate ultrasonic waves in liquid metals in a crucible strong enough to produce cavitation. Originally developed as a theoretical concept, this technique was tested experimentally in the treatment of liquid aluminium alloys. The computational model which couples fluid flow, heat transfer, electromagnetics and soundfield simulations was validated in aspects that are important for the process by the experimental results. These confirm the predicted free surface depression by the coil, induction driven flow leading to strong mixing and the presence of cavitation. It was found that acoustic resonant conditions are necessary to produce pressure waves of sufficient strength for cavitation, which means geometrical details and material sound properties of the setup become important for successful implementation. Using this technique, it has been possible to produce grain-refined samples of both pure aluminium and aluminium with grain refiner added.

Although the experiments presented so far are conducted in aluminium, since the technique is contactless, it should be equally applicable to high temperature or reactive metals, such as steels, nickel alloys and titanium, were the immersed sonotrode technique cannot be used. This is a subject of continuing research.

Author Contributions: K.P. is the overall project leader and Greenwich University grant holder and the main author, K.P. and V.B. developed the contactless sonotrode concept and V.B. produced the flow and heat transfer software and simulations. G.D. and C.T. developed the acoustic simulations, W.D.G. is the grant holder and PI for Birmingham University, A.D. carried out the experimental work and analysis together with W.D.G. All members contributed to the editing and provided material for the paper.

Funding: The authors acknowledge financial support from the ExoMet Project (EC contract FP7-NMP3-LA-2012-280421), and EPSRC grants EP/P034411/1, EP/R002037/1, EP/R000239/1.

Conflicts of Interest: The authors declare no conflict of interest.

References

1. Eskin, G.I.; Eskin, D.G. *Ultrasonic Treatment of Light Alloy Melts*, 2nd ed.; Georgy, I.E., Dmitry, G.E., Eds.; CRC Press: Boca Raton, FL, USA, 2017.
2. Meek, T.; Jian, X.; Xu, H.; Han, Q. *Ultrasonic Processing of Materials*; No. ORNL/TM-2005/125; University of Tennessee: Oak Ridge, TN, USA, 2006.
3. Tzanakis, I.; Xu, W.W.; Lebon, G.; Eskin, D.G.; Pericleous, K.; Lee, P.D. In situ synchrotron radiography and spectrum analysis of transient cavitation bubbles in molten aluminium alloy. *Phys. Procedia* **2015**, *70*, 841–845. [CrossRef]
4. Bojarevics, V.; Djambazov, G.S.; Pericleous, K.A. Contactless ultrasound generation in a crucible. *Met. Mater. Trans. A* **2015**, *46*, 2884–2892. [CrossRef]
5. Charles, V. Crystallization of aluminium alloys in the presence of cavitation phenomena induced by a vibrating electromagnetic pressure. *J. Cryst. Growth* **1996**, *158*, 118–127.
6. Ilmārs, G.; Gunter, G.; Andris, B. Contactless magnetic excitation of acoustic cavitation in liquid metals. *J. Appl. Phys.* **2015**, *117*, 204901.
7. Jarvis, D.; Pericleous, K.; Bojarevics, V.; Lehnert, C. Manufacturing of a metal component or a metal matrix composite component involving contactless induction of high-frequency vibrations. U.S. Patent No. 10,207,321, 19 February 2019.
8. Canuto, C.; Hussaini, M.Y.; Quarteroni, A.; Zang, T., Jr. *Spectral Methods in Fluid Dynamics*; Springer: Berlin, Germany, 1998.
9. Pericleous, K.; Bojarevics, V. Pseudo-spectral solutions for fluid flow and heat transfer in electro-metallurgical applications. *Prog. Comput. Fluid Dyn.* **2007**, *7*, 118–127. [CrossRef]
10. Djambazov, G.S.; Lai, C.-H.; Pericleous, K.A. Staggered-mesh computation for aerodynamic sound. *AIAAJ* **2000**, *38*, 16–21. [CrossRef]
11. Caflish, R.E.; Miksis, M.J.; Papanicolaou, G.C.; Ting, L. Effective equations for wave equations in bubbly fluids. *J. Fluid Mech.* **1985**, *153*, 259–273. [CrossRef]
12. Lebon, G.S.B.; Pericleous, K.A.; Tzanakis, I.; Eskin, D. A model of cavitation for the treatment of a moving liquid metal volume. *Int. J. Cast Met. Res.* **2016**, *29*, 324–330. [CrossRef]
13. Lebon, G.S.; Bruno, T.I.; Djambazov, G.; Pericleous, K.; Eskin, D.G. Numerical modelling of ultrasonic waves in a bubbly Newtonian liquid using a high-order acoustic cavitation model. *Ultrasonic Sonochem.* **2017**, *37*, 660–668. [CrossRef]
14. Wilcox, D.C. *Turbulence Modeling for CFD*, 2nd ed.; DCW Industries: La Cañada, CA, USA, 1998.
15. Tonry, C.E.H.; Djambazov, G.; Dybalska, A.; Bojarevics, V.; Griffiths, W.D.; Pericleous, K.A. Resonance from contactless ultrasound in alloy melts. In *Light Metals 2019*; Springer: Cham, Switzerland, 2019; pp. 1551–1559.
16. Tzanakis, I.; Hodnett, M.; Lebon, G.S.B.; Dezhkunov, N.; Eskin, D.G. Calibration and performance assessment of an innovative high-temperature cavitometer. *Sens. Actuators A: Phys.* **2016**, *240*, 57–69. [CrossRef]
17. Griffiths, W.D.; Beshay, Y.; Caden, A.J.; Fan, X.; Gargiuli, J.; Leadbeater, T.; Parker, D.J. The use of positron emission particle tracking (PEPT) to study the movement of inclusions in low melting point alloy castings. *Met. Mat. Trans. B* **2012**, *43B*, 370–378. [CrossRef]
18. Barigou, M. Particle tracking in opaque mixing systems: An overview of the capabilities of PET and PEPT. *Chem. Eng. Res. Des.* **2004**, *82*, 1258–1267. [CrossRef]
19. Leenov, D.; Kolin, A. Theory of Electromagnetophoresis. *J. Chem. Phys.* **1954**, *22*, 683–688. [CrossRef]

20. Manoylov, A.; Lebon, B.; Djambazov, G.; Pericleous, K. Coupling of acoustic cavitation with DEM-based particle solvers for modeling de-agglomeration of particle clusters in liquid metals. *MMTA* **2017**, *48*, 5616–5627. [CrossRef]
21. Leighton, T.G. Acoustic bubble detection. II. the detection of transient cavitation. *Environ. Eng.* **1995**, *8*, 16–25.
22. Pericleous, K.A.; Bojarevics, V.; Djambazov, G.I.; Dybalska, A.; Griffiths, W.; Tonry, C. The Contactless Electromagnetic Sonotrode. In *Shape Casting*; Springer: Cham, Switzerland, 2019; pp. 239–252.
23. Murty, B.S.; Kori, S.A.; Chakraborty, M. Grain refinement of aluminium and its alloys by heterogeneous nucleation and alloying. *Int. Mater. Rev.* **2002**, *47*, 3–29. [CrossRef]
24. Ager, P.; Iortsor, A.; Obotu, G.M. Behavior of aluminum alloy castings under different pouring temperatures and speeds. *Discovery* **2014**, *22*, 62–71.
25. Tzanakis, I.; Lebon, G.S.B.; Eskin, D.G.; Pericleous, K. Investigation of the factors influencing cavitation intensity during the ultrasonic treatment of molten aluminium. *Mater. Des.* **2016**, *90*, 979–983. [CrossRef]
26. Davis, J.R. *Aluminum and Aluminum Alloys*. ASM Specialty Handbook; ASM International, Metals Park: Novelty, OH, USA, 1993; pp. 201–210.
27. Quested, T.E.; Greer, A.L. Grain refinement of Al alloys: Mechanisms determining as-cast grain size in directional solidification. *Acta Mater.* **2005**, *53*, 4643–4653. [CrossRef]
28. Quested, T.E.; Greer, A.L. The effect of the size distribution of inoculant particles on as-cast grain size in aluminium alloys. *Acta Mater.* **2004**, *52*, 3859–3868. [CrossRef]
29. Easton, M.; StJohn, D. Grain refinement of aluminum alloys: Part I. the nucleant and solute paradigms—A review of the literature. *Met. Mater. Trans. A* **1999**, *30*, 1613–1623. [CrossRef]
30. Easton, M.A.; Stjohn, D.H. A model of grain refinement incorporating alloy constitution and potency of heterogeneous nucleant particles. *Acta Mater.* **2001**, *49*, 1867–1878. [CrossRef]
31. Pattnaik, A.B.; Das, S.; Bhushan Jha, B.; Prasanth, N. Effect of Al–5Ti–1B grain refiner on the microstructure, mechanical properties and acoustic emission characteristics of Al5052 aluminium alloy. *J. Mater. Res. Technol.* **2015**, *4*, 171–179. [CrossRef]
32. Djambazov, G.; Bojarevics, V.; Shevchenko, D.; Burnard, D.; Griffiths, W.; Pericleous, K. Sensitivity of Contactless Ultrasound Processing to Variations of the Free Surface of the Melt with Induction Heating. In *8th International Symposium on High-Temperature Metallurgical Processing*; Springer: Cham, Switzerland, 2017; pp. 289–298.

© 2019 by the authors. Licensee MDPI, Basel, Switzerland. This article is an open access article distributed under the terms and conditions of the Creative Commons Attribution (CC BY) license (http://creativecommons.org/licenses/by/4.0/).

Article

Effects of Ultrasonic Introduced by L-Shaped Ceramic Sonotrodes on Microstructure and Macro-Segregation of 15t AA2219 Aluminum Alloy Ingot

Tao Zeng [1] and YaJun Zhou [1,2,*]

1. College of Mechanical and Electrical Engineering, Central South University, Changsha 410083, China; taozg1203@163.com
2. National Key Laboratory of High Performance Complex Manufacturing, Central South University, Changsha 410083, China
* Correspondence: zhouyjun@csu.edu.cn; Tel.: +86-731-8887-7244

Received: 18 August 2019; Accepted: 23 September 2019; Published: 27 September 2019

Abstract: The effects of ultrasonic introduced by L-shaped sonotrodes made of high-temperature-resistant ceramic on the microstructure and macro-segregation of solidifying 15t AA2219 aluminum alloy ingots have been examined in the present study. The macroscopic morphology of the corrosion of the sonotrode has been observed. Grain refinement has been observed, the shape and size of the precipitated phase of the ingot were counted, and the degree of segregation along the transverse direction at 500 mm from the head of the ingot has been evaluated. The results reveal that the L-shaped ceramic ultrasonic introduction device can effectively avoid the erosion of high-temperature melt on the sonotrode and the heat radiation of the high-temperature heat flow to the transducer. Furthermore, the scanning electron microscope (SEM) and chemical composition detection results also indicate that the inter-dendritic micro-segregation of the equiaxed grains can be reduced, and the macro-segregation of the chemical composition of the ingot can be suppressed, and more homogeneous microstructures can be obtained when ultrasonic has been applied during solidification.

Keywords: ingot solidification; L-shaped ceramic sonotrode; grain refinement

---

1. Introduction

AA2219 aluminum alloy is a heat treatable reinforced aluminum alloy with Al, Cu, and Mn as its main alloying elements. Because of its good high- and low-temperature mechanical properties, formability, machinability, and welding properties, it has been widely used in the aerospace industry, especially as the main material for the new generation of launch vehicle propellant tanks [1–5]. The original billet of the tank is a large-sized aluminum alloy ingot, which is difficult to form. It adopts the traditional casting method and is prone to defects such as loose shrinkage, coarse grain, serious segregation, and cracking of the structure, which extremely affect its performance and lead to its unstable production in batches [6,7].

Li Xiaoqian et al. [8,9] conducted a large number of casting experiments on 2XXX and 7XXX aluminum alloys using a straight-rod titanium alloy ultrasonic introduction device. The results showed that ultrasonic vibration treatment can improve the solidification structure of aluminum alloy and it can also degas, reduce microscopic loosening, refine grains, and improve the effect of improve micro-segregation. Wang Jianwu et al. [10] applied an ultrasonic wave with a power of 179 W, a frequency of 19 kHz, and a sound wave intensity of 153.33 W/cm^2 to a 7050 aluminum alloy melt, with a treatment time of about 3 min. The experimental results showed that the microstructure of an ingot that has not been treated with ultrasonic is coarse dendrites with precipitated phase particles.

After applying ultrasonic waves, both grains and the second phase particles of the ingot are refined, and the average grain size is reduced from 80 µm to 40 µm. Wang G et al. [11] studied the effect of the temperature range of melt when applying ultrasonic during solidification of Al-2Cu alloy on grain structure and cooling behavior. The studies show that when introducing ultrasound into the melt, grain refinement requires an appropriate amount of liquid metal overheating or sufficient preheating of the sonotrode.

Haghayeghi R et al. [12] introduced ultrasonic vibration to a melt before AA7075 high pressure die casting and found that, due to cavitation, the porosity decreased by 5%, the tensile strength and yield strength increased to 590 MPa and 502 MPa respectively, and the elongation increased to 18%. They pointed out that ultrasonic casting is feasible for industrial-scale applications.

There are many studies on the ultrasonic casting of aluminum alloy, but most of them are limited to laboratory research, and the ultrasonic conduction tools used are straight-rod sonotrodes. When applying ultrasonic waves, the heat radiated by the melt will seriously affect the stability of the ultrasonic vibration system. Moreover, the titanium alloy sonotrode is prone to erosion during the casting process [13], and it is impossible to stably conduct ultrasonic waves into the molten metal for a long time. Therefore, ultrasonic-assisted casting technology has not been applied in the industry so far. In order to solve this problem, an L-shaped ceramic ultrasonic introduction device was used to introduce ultrasonic waves into the melt in the 2219 aluminum alloy casting process. The L-shaped ceramic ultrasonic introduction device can avoid the direct heat radiation of the high-temperature melt to the transducer. The ceramic sonotrode can also overcome the problem of cavitation erosion caused by the high-temperature aluminum melt and improve the stability of the ultrasonic vibration system. Through experiments without ultrasonic casting, straight-rod ultrasonic-assisted casting, and L-shaped ultrasonic-assisted casting, the effects of cavitation melt processing by an L-shaped sonotrode made of high-temperature-resistant ceramic on solidifying 15t AA2219 aluminum alloy ingots were discussed.

2. Materials and Methods

2.1. Casting

Without ultrasonic casting, the 2219 aluminum alloy was proportioned according to GB/T3190-2008 standard. 15t commercial purity aluminum was put into a 20t-capacity melting furnace, heated and melted, which was then fully stirred by the electromagnetic stirrer at the bottom of the furnace. The following step was slag removal, after which metal additives like Cu and Mn were added into the aluminum melt. The refined product was then taken to degas and remove slag. After another slagging, the direct reading spectrometer was used to analyze the components, the results of which are shown in Table 1. After all components reached the standard, Al–Ti–B refiner was added. Then the melt was sent to the hot top crystallizer along a channel for casting. The casting parameters are listed in Table 2. An ingot (Named NO-UTS ingot) with a size of Φ650 mm × 6300 mm was produced and moved to the heat treatment furnace for homogenization heat treatment at the temperature of 540 °C for 60 h.

Table 1. Chemical composition of 2219 aluminum alloy (wt. %).

Chemical Composition	Cu	Mn	Fe	Ti	V	Zr	Si	Al
NO-UTS	5.9	0.36	0.018	0.04	0.07	0.11	0.024	Bal
I-UTS	5.91	0.35	0.02	0.04	0.07	0.11	0.025	Bal
L-UTS	5.9	0.34	0.019	0.04	0.06	0.11	0.023	Bal

Table 2. Casting parameters of aluminum alloy ingot.

Specification/ (mm × mm)	Pouring Temperature/°C	Cooling Temperature/°C	Speed of Water Flow/(L/min)	Speed of Introducing Ingot/(mm/min)
Φ650 × 6200	696	29.1	321	24

With ultrasonic casting, under the condition that other technological conditions are as constant as possible after casting becomes stable, the ultrasonic wave was directly introduced from the hot top. Then the straight-rod titanium alloy sonotrode and L-shaped ceramic sonotrode made by our group were used to introduce an ultrasonic wave (the structural sketches of the two ultrasonic rods are shown in Figure 1). Two different sonotrodes with a size of Φ50 mm × 185mm were inserted into the liquid surface below about 50 mm. The piezoceramic transducer was used to convert high frequency current from the generator to mechanical oscillations of the same frequency. The ultrasonic power was 800 W, the ultrasonic frequency was 23 ± 0.2 KHz, and the amplitude was 15 μm. In the final stage of casting, when the melt level dropped, the ultrasonic vibration stopped and the ultrasonic vibration system was removed. Two Φ650 mm × 6300 mm ingots (I-UTS ingot and L-UTS, separately) were cast separately (the casting device is shown in Figure 2), and the cast large ingots were transferred into the heat-treatment furnace and homogenized through the same heat-treatment process.

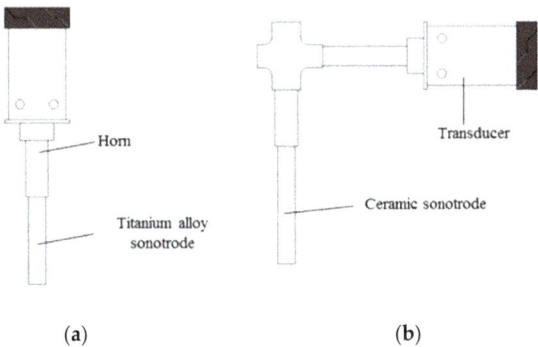

Figure 1. Structure of ultrasonic-introducing sonotrodes: (**a**) straight-rod sonotrode made of titanium alloy; (**b**) L-shaped sonotrode made of ceramic.

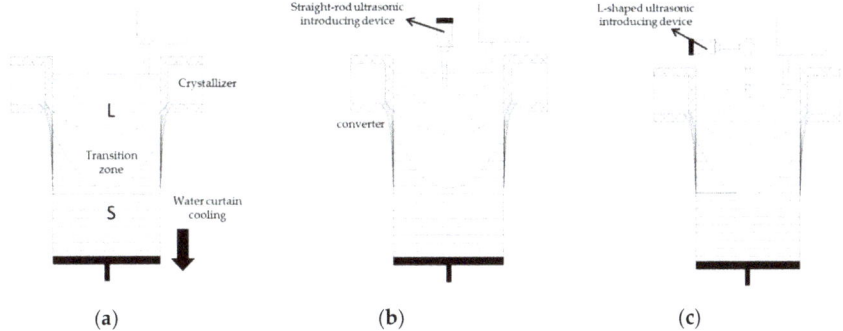

Figure 2. Different casting processes: (**a**) semi-continuous casting without ultrasonic treatment; (**b**) semi-continuous casting with the straight ultrasonic-introducing rod; (**c**) semi-continuous casting with L-shaped ultrasonic-introducing rod.

2.2. Sampling

After the 10 mm thick oxide inclusion layer was removed from the surface of three round ingots, the samples with a specification of Φ630 mm × 25 mm were cut at 500 mm from the top of the ingot. The sample was cut along the radial diameter, one part for macrostructure analysis, and the other part for microstructure analysis and chemical composition analysis. For the latter, we needed to take smaller samples at positions of 0R, 1/4R, 1/2R, 3/4R, R. The sampling locations are shown in Figure 3.

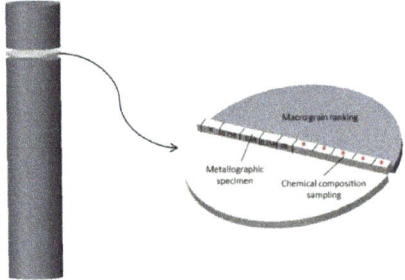

Figure 3. Schematic diagram.

2.3. Characterization

Firstly, the erosion morphology of the sonotrode of the ultrasonic-introducing device was observed after ultrasonic treatment. Then, according to the GB/T 3246.2-2000 standard, the sample taken from the ingot was polished with sandpaper, washed with alcohol, and dried. Then, it received macro corrosion and macro grain ranking. The corrodent was prepared with 10 mL of HF with 42% concentration, 5 mL of HCl with 36% concentration, 5 mL of HNO_3 with 65% concentration, and 380 mL of water. According to the GB/T 3246.2-2000 standard, the metallographic specimen was ground and polished on an MP-2B grinding and polishing machine (Weiyi Test Equipment Manufacturing Corporation, Laizhou, China), washed with clear water, and etched with the above corrodent for 60 s. Then the metallurgical structure of the ingot was observed through an OLYCLA-DSX500(OLYMPUS Corporation, Tokyo, Japan) metallurgical microscope and the morphology, size, and distribution of its precipitation phase were observed through a Phenom automatic table scanning electron microscope (Phenom-world BV, Eindhoven, Holland). In addition, the amount of alloying element was analyzed by EDS (energy dispersion spectrometer). An Inductively Coupled Plasma Optical Emission Spectrometer (ICP-OES, Shimadzu Corporation, Tokyo, Japan) was also used to test the chemical components of different positions.

3. Results and Discussions

3.1. Anti-Corrosion Property of L-Shaped Sonotrode

The morphology of the straight-rod and L-shape sonotrode after 24 h of experiment is shown in Figure 4. It can be seen that the straight-rod titanium alloy sonotrode is seriously eroded, while the L-shaped sonotrode is basically not eroded. The titanium alloy sonotrode selected in this experiment is made of titanium alloy TC_4, which can withstand a temperature of 1700 °C and conduct ultrasonic waves efficiently, and has the advantages of excellent corrosion resistance, low density, high specific strength, and good toughness. However, in the process of introducing ultrasound, it was slowly eroded, mainly due to the formation of alloy between the titanium alloy sonotrode and metal melt [14,15], indicating that although a titanium alloy sonotrode can withstand high temperature, it cannot withstand the melting corrosion effect of metal melt. The L-shaped ultrasonic introduction device is composed of a transducer, an L-shaped combined horn, and an upright ceramic sonotrode [16] (Figure 1b). The L-shaped combined horn is composed of a first-stage horizontal horn and a second-stage vertical horn, the first-stage horn is a straight rod that propagates longitudinal mechanical vibration, and the second-stage horn is a straight rod that propagates lateral mechanical vibrations (as shown in Figure 1b). The material used for the L-shaped combined horn is titanium alloy. The material used for the sonotrode is a special ceramic with good cold and thermal shock resistance and high strength at high temperature, and a melting point that is higher than 1900 °C. It is not oxidized below 1200 °C, the protective film formed at 1200–1600 °C can prevent further oxidation, it can resist vibration and thermal shock, and cannot form alloys with metal melts [17]. Therefore, the sonotrode will not be eroded by the melt. In addition, the L-shaped combination keeps the ultrasonic transducer away from the high-temperature metal melt during operation, effectively avoiding the direct thermal radiation of the high-temperature heat flow to the transducer, thus prolonging the life of the ultrasonic transducer while the energy of the ultrasonic is continuously inserted into the metal melt.

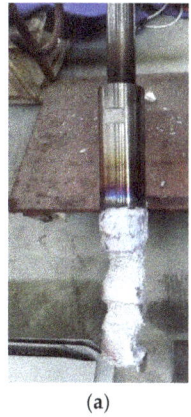
(a)

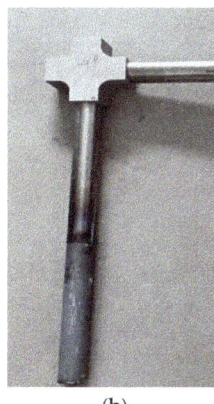
(b)

Figure 4. Morphology of sonotrodes: (a) straight-rod titanium alloy sonotrode; (b) L-shaped ceramic sonotrode.

The performance of sonotrode plays a key role in the effect of ultrasonic treatment, and the performance of sonotrode mainly depends on two main output parameters: resonance frequency and displacement amplitude. In order to compare the performance of straight-rod sonotrode and L-shaped sonotrode, it is necessary to measure the resonant frequency and output displacement amplitude of two ultrasonic introduction systems. The output amplitude of the ultrasonic vibration system used in this experiment is relatively small, so the Laser Doppler Velocimeter (LDV) is used to measure it. As shown in Table 3, the maximum amplitude and resonance frequency at the end face of the L-shaped sonotrode are 15.21 μm and 23.08 KHz, respectively. The maximum amplitude and resonance frequency at the end face of the straight-rod sonotrode are 11.67 μm and 23.13 KHz respectively. It can be seen that the ultrasonic introduction effect of L-shaped sonotrode is better than that of straight-rod.

Table 3. Actual output amplitude and resonance frequency of different ultrasonic vibration systems.

Type	Amplitude/μm	Frequency/KHz
I-shaped	15.21	23.08
L-shaped	11.67	23.13

3.2. Effects on Macrostructure and Microstructure

Figure 5 displays the macro grain grade diagram of the 2219 aluminum alloy ingots under three different casting processes. The macro grain size grade of the NO-UTS ingot was 1–1+ grade in the edge, 1.5–2.5 grade in the middle, and 2.5–2 grade in the center. The macro grain size grade of the I-UTS ingot was 1–1+ grade at the edge, 1.5 grade in the middle, and 1.5 grade in the center. The macro grain size grade of L-UTS ingot was 1–1+ grade at the edge, 1–2 grade in the middle, and 1.5 grade in the center. The macro grain size of the three ingots all shared the characterization of being smallest at the edge, largest in the middle, and smaller in the center. As for the smallest size at the edge, the reason is that when the liquid aluminum flows into the crystallizer, the temperature of the die wall is very low due to the action of cooling water. As the undercooling of the edge area increases, resulting in a large number of nucleation, the die wall can be used as the starting point of nucleation to form a dense and fine equiaxed crystal zone. At the same time as the formation of the equiaxed crystal zone, the temperature of the equiaxed crystal zone increases, and the latent crystal of the equiaxed crystal zone releases the latent heat of crystallization. The cooling rate of the remaining liquid is reduced, and the degree of undercooling is reduced, so that the grain size of the middle region is larger; the grain size in the center gets smaller than in the middle because the flow of melt brings some un-melted impurities or broken dendritic crystals to the center of the ingot and heterogeneous nucleation forms thereby, though the cooling rate in the central region is slower and the temperature difference is smaller [18,19]. The grain size grade of the ingot without ultrasonic treatment was large, and the grain size grade of each position was quite different. After ultrasonic treatment with L-shaped sonotrode, the macroscopic grain size grades in the middle and the center were significantly reduced, which can effectively refine and homogenize the grain structure, thus improving the macro-grain structure of the ingot, with a better grain refinement effect than that of the straight-rod sonotrode.

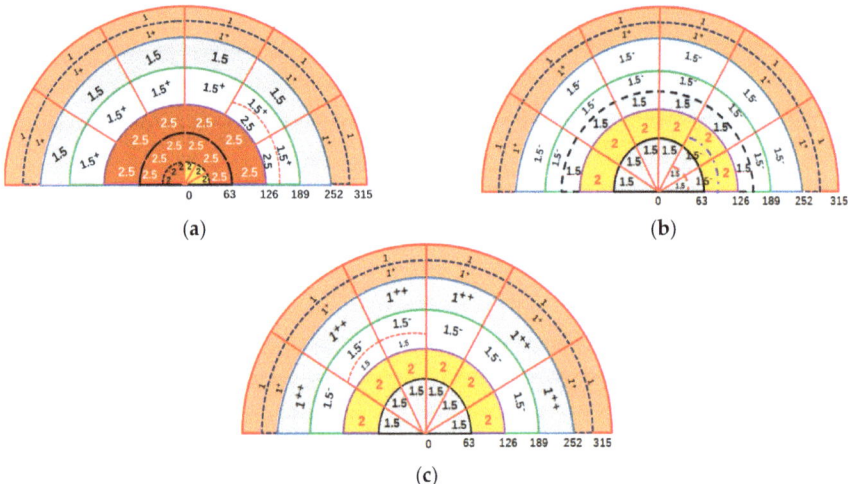

Figure 5. Grain grade diagram of ingots: (a) NO-UTS; (b) I-UTS; (c) L-UTS.

Microstructure in the radial regions of ingots is shown in Figure 6. It can be seen from the figure that the grain size of the NO-UTS ingot is coarse and uneven, and the grain structure has been obviously refined by ultrasonic treatment. The grain size of L-UTS ingot is refined remarkably in all radial positions, and the grain size distribution is more homogenized. Compared with the I-UTS ingot, the grain structure is finer in the regions of 0.5R–0.75R. The grain size of all ingots increased first and then decreased along the radial direction, which is consistent with the macroscopic grain rating of the ingot in the above figure.

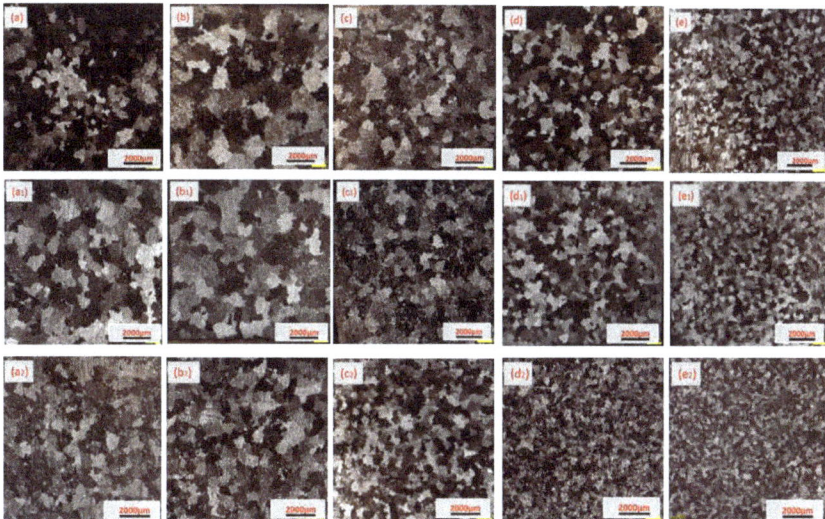

Figure 6. In the regions of the three ingots: (**a–e**) positions of 0R, 0.25R, 0.5R, 0.75R, R of the NO-UTS ingot; (a_1–e_1) positions of 0R, 0.25R, 0.5R, 0.75R R of I-UTS ingot; (a_2–e_2) positions of 0R, 0.25R, 0.5R, 0.75R, R of the L-UTS ingot.

The average grain sizes in radial direction of the three ingots are shown in Figure 7. Compared with the NO-UTS ingot, the grain size of the L-UTS ingot reduced from 547–1100 μm to 464–874 μm, its average size decreased from 864.2 μm to 750.1 μm, and SD (standard deviation) decreased from 195.23 to 151.06. Compared with the I-UTS ingot, the L-UTS ingot had a better refining effect and produced a finer ingot with a grain size of 451–771 μm, an average grain size of 632.74 μm and SD of 112.8.

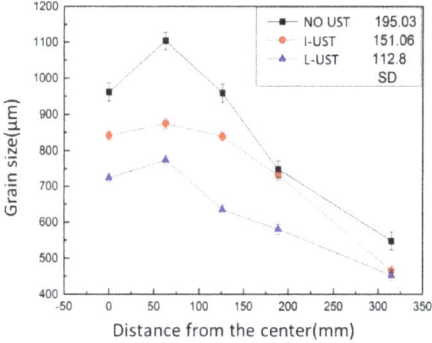

Figure 7. Of grains in the radial direction of the three ingots.

The cavitation and sound flow effects of ultrasonic during the ultrasonic casting process are the main factors affecting the solidification process of aluminum melt. The application of ultrasonic vibration in the aluminum melt can significantly increase the vibration frequency and energy of nucleus and produce a large number of cavitation bubbles in the melt. When the cavitation bubbles are resonated by ultrasound, the vibration frequency and energy of the nucleus will be significantly increased. The nucleus vibration can inhibit the growth of grains, which is conducive to the formation of equiaxed crystals and dendrites. The dendrites are also broken by vibration to form new crystal nuclei, so that the nucleation rate is improved and the crystal grains are refined. In addition, when cavitation bubbles collapse due to sound intensity, the strong shockwave will also break the coarse dendrites, and, under the stirring action of the sound flow, the broken dendrites disperse into the melt and become effective nucleation particles, so that the nucleation rate is improved and the grain size distribution is more homogenized. Finally, the strong stirring effect of ultrasound can increase the particle diffusion rate, make the solute in the melt more homogenized, and reduce the component undercooling in the front of crystallization. Small component undercooling can increase the liquidus temperature of the aluminum solution, increasing the effective undercooling and the nucleation rate, and significantly refining the grain size [20–23].

The cavitation threshold is a quantity indicating the difficulty of cavitation in the melt. The ultrasound cavitation threshold is related to many factors. In the same melt, it is closely related to the frequency, waveform and waveform parameters of ultrasound, and it increases with an increase of ultrasonic power. In the L-shaped ultrasonic introduction device, the primary and secondary titanium alloy horns are combined in an L shape, which converts the longitudinally transmitted sine wave into a laterally transmitted distortion wave, so that the mechanical vibration and sound flow effect of the ultrasonic wave in the melt are strengthened [24]. As a result, the threshold of cavitation is smaller, cavitation can occur more easily, and the grain size is smaller and more uniform than that of a straight-rod ultrasonic introduction device.

3.3. Effects on the Secondary Phase

Semi-continuous casting produces large casting stress which should be eliminated through homogenization annealing to prevent cracking in subsequent mechanical processing [25]. In addition, it can promote the re-melting of the low-melting eutectic phase in the alloy to some extent, eliminate or inhibit the inhomogeneity of the microstructure and chemical composition in the grain, and increase the solid solubility of the alloying elements in the matrix so the strength of the alloy is improved [26]. The SEM images of different positions of the three ingots are shown in Figure 8. The gray area is the α-Al matrix of the ingot and the white area is the precipitation phase. The morphology of precipitated phases in ingots is needle-shaped, and granular and spherical, needle-like precipitates were most common in the NO-UTS ingot, followed by the I-UTS ingot, and least in the L-UTS ingot (as shown in Figure 9). The needle-like precipitates tend to have sharp edges and corners, which tend to cause stress concentration at the tip, which is not conducive to the plastic deformation of the matrix. The symmetrical, smooth spheroidal precipitation has a small splitting effect on the matrix, which is beneficial to the uniform plastic deformation of the matrix. Therefore, the L-UTS ingot is most conducive to the subsequent processes of the ingot. In addition, the size and distribution of the precipitated phase of the NO-UTS ingot are uneven as the precipitated phases are larger from the center to the position of 1/2R while becoming small at the edge, and an enrichment phenomenon occurs in some areas. This problem is improved partly by I-UTS casting while the precipitated phase at the position of 1/2R is still large and the enrichment phenomenon occurs. However, the precipitated phases are not only generally refined but also become more evenly distributed in the L-UTS ingot.

The distribution of the secondary phase at the position of 0.75R in the crystal of three kinds of ingots is shown in Figure 10. For the NO-UTS ingot, the secondary phases in the crystal were mainly block-shaped and needle-shaped, with an average size of 4.1 µm. For the I-UTS ingot, the secondary phases were mainly block-shaped and dot-shaped, with an average size of 3.15 µm. For the L-UTS ingot, the secondary phases were mainly dot-shaped, with an average size of 2.38 µm. It can be seen that the secondary phase of the whole section was obviously refined by applying ultrasonic, and the agglomeration and growth of the secondary phase were effectively controlled. The energy spectrum analysis of the secondary phase in the crystal shows that the atomic number ratio of Al to Cu is 2:1 (as shown in Table 4), which indicates the secondary phase was Al_2Cu. The effect of high-speed acoustic streaming generated by ultrasonic could stir the metal melt to some extent, thus promoting the diffusion and solid solution of alloy elements, reducing the segregation of alloy elements, and strengthening the matrix [27].

Table 4. The mole fraction of the second phase in different ingot crystals (%).

Testing Point	Al	Cu
1	64.33	35.67
2	63.17	36.83
3	64.02	35.98

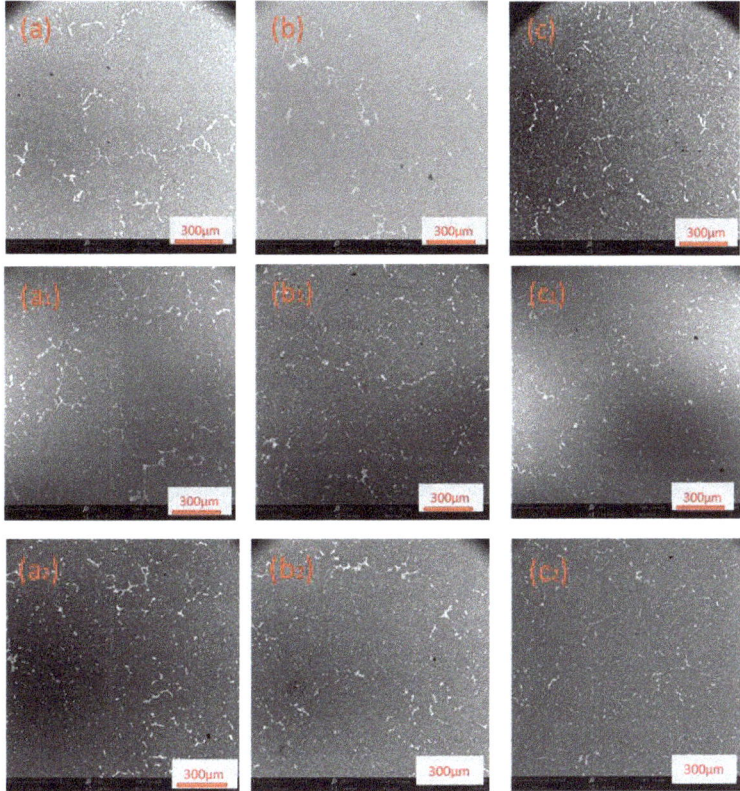

Figure 8. Phases of ingots under different processing parameters: (a–c) positions of 0R, 0.5R, R of NO-UTS ingot; (a_1–c_1) positions of 0R, 0.5R, R of I-UTS ingot; (a_2–c_2) positions of 0R, 0.5R, R of the L-UTS ingot.

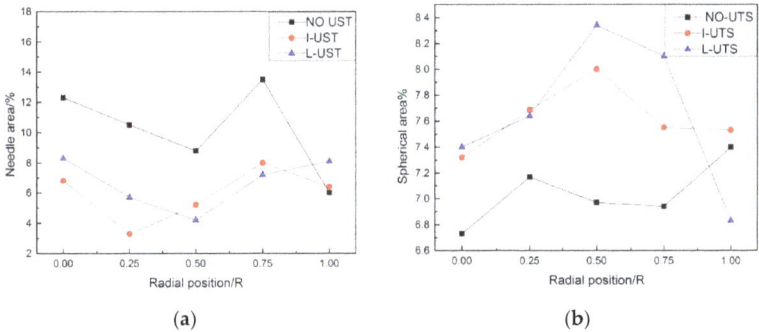

Figure 9. Of precipitated phase areas in different shapes: (a) needle area; (b) spherical area.

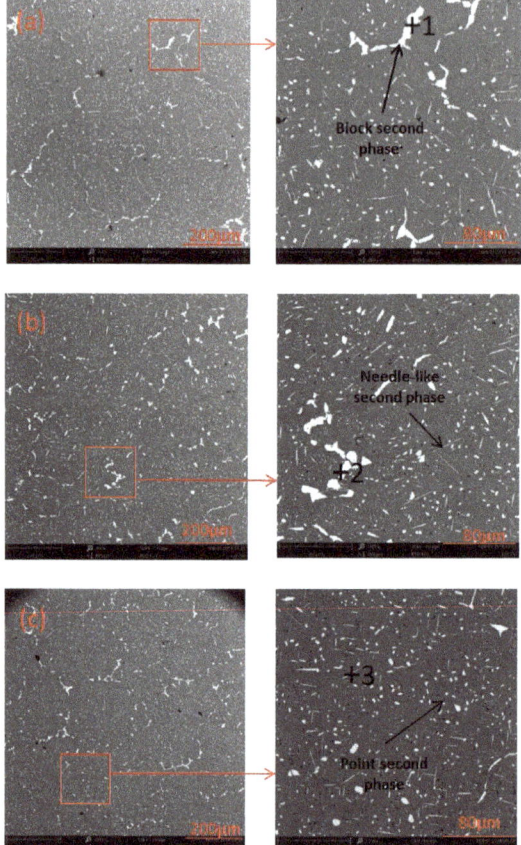

Figure 10. Secondary phases of ingots at the position of 0.75 R under different processing parameters: (a) NO-UTS; (b) I-UTS; (c) L-UTS.

3.4. Effects on Macro-Segregation

The degree of macro-segregation of the solute can be characterized by segregation rate S, which is obtained by the difference between ΔC_{max} (the maximum relative segregation rate) and ΔC_{min} (the minimum relative segregation rate), wherein the relative segregation rate is $\Delta C = (C_i - C_0)/C_0$, C_i the concentration percentage of the element measured at each position, and C_0 is the initial concentration percentage of the element. $\Delta C > 0$ represents positive segregation while $\Delta C < 0$ represents negative segregation [28]. We sampled by drilling five holes in a radial direction (Figure 3) and carried out a chemical composition test so as to know the solute element distribution along the radial direction on the cross-section of the round ingot. The distribution diagram of the relative segregation rate of Cu and Ti in 2219 aluminum alloy is shown in Figure 11, taking the radial direction as the x-axis and the relative segregation rate ΔC as the y-axis.

There was an obvious negative segregation at the edge and positive segregation at the center in all aluminum ingots in the three trials. This phenomenon is caused by the solidification rule of 2219 aluminum alloy. The relative segregation rate of Cu of the NO-UTS ingot fluctuates greatly, that is, the concentration of Cu fluctuates greatly along the radial direction and a great difference of concentration between the center and the edge could be seen. Although the uneven distribution of Cu along the radial direction was improved after ultrasonic is applied, the negative segregation of

Cu reduced due to a higher concentration at the edge while the positive segregation of Cu decreased due to lower concentration at the center. The segregation rates of Cu in the I-UTS ingot and L-UTS ingot were quite the same. It is calculated that S = 0.127 in NO-UTS, S = 0.057 in the I-UTS ingot, and S = 0.058 in the L-UTS ingot, and the segregation rate of Cu was decreased by 54.3% with L-shaped ultrasonic casting. These results illustrate that ultrasonic-assisted casting can significantly reduce the macro-segregation of Cu in 2219 large aluminum alloy ingot.

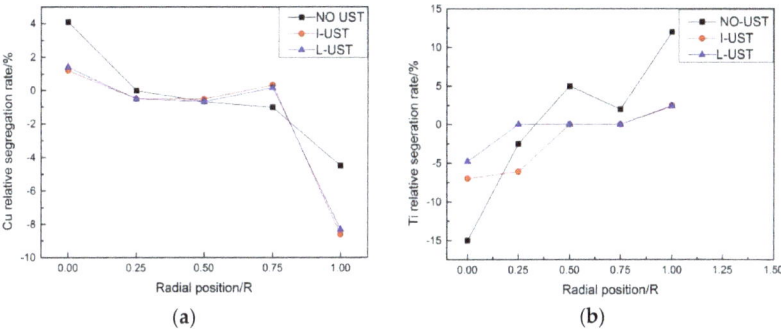

Figure 11. Segregation rate: (a) Cu; (b) Ti.

AlTiB wire alloy is a common grain refiner for aluminum alloy casting, and its refinement mechanism is explained by the double nucleation theory, which asserts that TiB_2 particles with thin $TiAl_3$ layers act as nuclei in crystallization. Thus, the grain refinement properties of solid aluminum alloys depend largely on the size and shape of TiB_2 particles and $TiAl_3$ phases in their microstructures [29,30]. From the relative segregation rate of Ti element at different positions along the radial direction (Figure 11), it can be seen that the Ti element in the NO-UTS ingot had a larger positive segregation at the edge and larger negative segregation at the center. This is because Ti element will undergo a peritectic reaction with aluminum, and the segregation rate shows a tendency opposite to that of Cu element. The segregation rates and fluctuation significantly reduced after the melting of casting aluminum were treated by ultrasound. This is because the local high temperature and high pressure generated by the ultrasonic accelerates the dissolution of the coarse $TiAl_3$ particles, and the ultrasonic flow agitation increases the diffusion distance of the solute Ti and uniformly distributes it in the aluminum solution. The negative pressure generated by the cavitation effect can increase the surface energy of TiB_2 particles by removing gases, impurities, and oxides on the particle surface, thus improving the wettability between TiB_2 particles and aluminum melt [31]. The local high temperature and high pressure produced by cavitation effects also reduce the surface tension of aluminum melt, which can further improve the wettability between particles and melt. Meanwhile, the stirring by acoustic streaming can also boost a more homogenized distribution of TiB_2 particles in the melt [32,33]. Thus, the thin $TiAl_3$ layer on the surface of TiB_2 particles is formed and a more homogenized and finer grain structure is obtained [34,35].

4. Conclusions

1. The L-shaped ceramic ultrasonic introduction device can effectively avoid the erosion of high-temperature melt on sonotrode and the heat radiation of the high-temperature heat flow to the transducer.
2. The ultrasound introduced by the L-shaped ultrasonic introduction device made of ceramic can refine grains better and make their size and macro distribution more homogenized than the straight-rod ultrasonic introduction device due to the stronger mechanical vibration and sound flow effect in the melt.

3. L-shaped ultrasonic-assisted casting can produce more dot-shaped and globular precipitation phases conducive to plastic deformation of the matrix and make the distribution of precipitated phases more homogenized with less concentration than straight-rod ultrasonic-assisted casting.
4. Both ultrasonic-assisted castings can decrease macro-segregation of the chemical composition in the AA2219 aluminum alloy ingot, causing the concentration of Cu and Ti to distribute more evenly in the whole section.

Author Contributions: Writing—original draft preparation, T.Z.; writing—review and editing, Y.Z.

Funding: This research received no external funding.

Acknowledgments: The authors would like to acknowledge the financial assistance provided by the Major State Basic Research Development Program of China (No. 2014CB046702).

Conflicts of Interest: The authors declare no conflict of interest.

References

1. Wang, H.; Yi, Y.; Huang, S. Influence of Pre-Deformation and Subsequent Ageing on the Hardening Behavior and Microstructure of 2219 Aluminum Alloy Forgings. *J. Alloy. Compd.* **2016**, *685*, 941–948. [CrossRef]
2. Lam, A.C.L.; Shi, Z.; Yang, H.; Wan, L.; Davies, C.M.; Lin, J.; Zhou, S. Creep-Age Forming AA2219 Plates with Different Stiffener Designs and Pre-Form Age Conditions: Experimental and Finite Element Studies. *J. Mater. Process. Technol.* **2015**, *219*, 155–163. [CrossRef]
3. An, L.; Cai, Y.; Liu, W.; Yuan, S.; Zhu, S.; Meng, F. Effect of Pre-Deformation on Microstructure and Mechanical Properties of 2219 Aluminum Alloy Sheet by Thermomechanical Treatment. *Trans. Nonferrous Met. Soc. China* **2012**, *22*, 370–375. [CrossRef]
4. Zhang, D.; Wang, G.; Wu, A.; Zhao, Y.; Li, Q.; Liu, X.; Meng, D.; Song, J.; Zhang, Z. Study on the Inconsistency in Mechanical Properties of 2219 Aluminum Alloy TIG-welded Joints. *J. Alloy. Compd.* **2019**, *777*, 1044–1053. [CrossRef]
5. Zhang, L.; Li, X.; Li, R.; Jiang, R.; Zhang, L. Effects of High-Intensity Ultrasound on the Microstructures and Mechanical Properties of Ultra-Large 2219 Al Alloy Ingot. *Mater. Sci. Eng. A* **2019**, *763*, 138154. [CrossRef]
6. Cao, H.; Hao, M.; Shen, C.; Liang, P. The Influence of Different Vacuum Degree on the Porosity and Mechanical Properties of Aluminum Die Casting. *Vacuum* **2017**, *146*, 278–281. [CrossRef]
7. Paramatmuni, R.K.; Chang, K.; Kang, B.S.; Liu, X. Evaluation of Cracking Resistance of DC Casting High Strength Aluminum Ingots. *Mater. Sci. Eng. A* **2004**, *379*, 293–301. [CrossRef]
8. Ni, Z.H.; Li, X.Q.; Zhang, M. Effect of ultrasonic external field on solidification microstructure of 7050 aluminum alloy ingot. *Mater. Sci. Eng. Powder Metall.* **2011**, *16*, 341–348.
9. Zhang, M.; Li, X.Q.; Li, Z.H. Microscopic looseness in 7050 alloy under high power ultrasonic field. *China Foundry* **2011**, *60*, 136–139.
10. Wang, J.W.; Xiong, Y.X. Effect of ultrasonic on the microstructure and aging characteristics of 7050 aluminum alloy. *Spec. Cast. Nonferrous Alloy* **2012**, *32*, 1086–1090.
11. Wang, G.; Dargusch, M.S.; Qian, M.; Eskin, D.G.; StJohn, D.H. The Role of Ultrasonic Treatment in Refining the As-Cast Grain Structure During the Solidification of an Al–2Cu Alloy. *J. Cryst. Growth* **2014**, *408*, 119–124. [CrossRef]
12. Haghayeghi, R.; Heydari, A.; Kapranos, P. The Effect of Ultrasonic Vibrations Prior to High Pressure Die-Casting of AA7075. *Mater. Lett.* **2015**, *153*, 175–178. [CrossRef]
13. Tian, Y.; Liu, Z.; Li, X.; Zhang, L.; Li, R.; Jiang, R.; Dong, F. The Cavitation Erosion of Ultrasonic Sonotrode During Large-Scale Metallic Casting: Experiment and Simulation. *Ultrason. Sonochem.* **2018**, *43*, 29–37. [CrossRef] [PubMed]
14. Barbosa, J.; Puga, H. Ultrasonic Melt Processing in the Low Pressure Investment Casting of Al Alloys. *J. Mater. Process. Technol.* **2017**, *244*, 150–156. [CrossRef]
15. Lebon, G.S.B.; Tzanakis, I.; Pericleous, K.; Eskin, D. Experimental and Numerical Investigation of Acoustic Pressures in Different Liquids. *Ultrason. Sonochem.* **2018**, *42*, 411–421. [CrossRef] [PubMed]
16. Shi, C.; Shen, K.; Mao, D.H.; Liang, G.; Li, F. Research Progress on Ceramic Tool Rod for Ultrasound Treatment of Metal Melt. *Hot Work. Technol.* **2018**, *2018*, 5–9.

17. Liang, G.; Shi, C.; Zhou, Y.; Mao, D. Numerical Simulation and Experimental Study of an Ultrasonic Waveguide for Ultrasonic Casting of 35CrMo Steel. *J. Iron Steel Res. (Int.)* **2016**, *23*, 772–777. [CrossRef]
18. Guo, C.W.; Li, J.J.; Yuan, M.; Wang, J.C. Growth Behaviors and Forced Modulation Characteristics of Dendritic Sidebranches in Directional Solidification. *Phys. Sin.* **2015**, *64*. [CrossRef]
19. Liang, G.; Chen, S.; Yajun, Z.; Daheng, M. Effect of Ultrasonic Treatment on the Solidification Microstructure of Die-Cast 35CrMo Steel. *Met. Open Access Metall. J.* **2016**, *6*, 260. [CrossRef]
20. Liu, X.; Osawa, Y.; Takamori, S.; Mukai, T. Grain Refinement of AZ91 Alloy by Introducing. Ultrasonic Vibration During Solidification. *Mater. Lett.* **2008**, *62*, 2872–2875. [CrossRef]
21. Flannigan, D.J.; Suslick, K.S. Plasma Formation and Temperature Measurement During Single-Bubble Cavitation. *Nature* **2005**, *434*, 52–55. [CrossRef]
22. Shi, C.; Shen, K. Twin-Roll Casting 8011 Aluminum Alloy Strips Under Ultrasonic Energy Field. *Int. J. Lightweight Mater. Manuf.* **2018**, *1*, 108–114.
23. Abramov, V.; Abramov, O.; Bulgakov, V.; Sommer, F. Solidification of Aluminum Alloys Under Ultrasonic Irradiation Using Water-Cooled Resonator. *Mater. Lett.* **1998**, *37*, 27–34. [CrossRef]
24. Zhang, M. Effect of Longitudinal and Transverse Vibration of Ultrasonic Radiator on Cavitation and Refining Region of Aluminum Melt. Master's Thesis, Central South University, Changsha, China, 2014.
25. Shi, Y.J.; Pan, Q.L.; Li, M.J.; Liu, Z.M.; Huang, Z.Q. Microstructural Evolution During Homogenization of DC Cast 7085 Aluminum Alloy. *Trans. Nonferrous Met. Soc. China* **2015**, *25*, 3560–3568. [CrossRef]
26. Guo, X.; Jiang, R.; Xiaoqian, L.I.; Ruiqing, L.I.; Peng, B. Effect of Aluminum Alloy Melt Treated with Ultrasound on Microstructure and Segregation of Large-Scale Ingot in Semi-Continuous Casting. *Hot Work. Technol.* **2016**, *45*, 25–29.
27. Chen, D.; Li, X.; Guo, D.; Cui, Y.; Jiang, R. Microstructure and Macrosegregation of 7050 Alloy Large Flat Billet in Ultrasonic Semi-Continuous Casting. *Spec. Cast. Nonferrous Alloy* **2012**, *32*, 462–465.
28. Zhang, L.; Li, X.; Li, R.; Jiang, R.; Zhang, L. Effect of Ultrasonication On the Microstructure and Macrosegregation of a Large 2219 Aluminum Ingot. *Chin. J. Eng.* **2017**, *39*, 1347–1354.
29. Li, X. Effect of High Intensitive Ultrasound on AlTiC Refiner. *J. Huazhong Univ. Sci. Technol.* **2013**, *4*, 11–15.
30. Tzanakis, I.; Xu, W.W.; Eskin, D.G.; Lee, P.D.; Kotsovinos, N. In Situ Observation and Analysis of Ultrasonic Capillary Effect in Molten Aluminium. *Ultrason. Sonochem.* **2015**, *27*, 72–80. [CrossRef]
31. Lebon, G.S.B.; Tzanakis, I.; Pericleous, K.; Eskin, D.; Grant, P.S. Ultrasonic Liquid Metal Processing: The Essential Role of Cavitation Bubbles in Controlling Acoustic Streaming. *Ultrason. Sonochem.* **2019**, *55*, 243–255. [CrossRef]
32. Lebon, G.S.B.; Salloum-Abou-Jaoude, G.; Eskin, D.; Tzanakis, I.; Pericleous, K.; Jarry, P. Numerical Modelling of Acoustic Streaming During the Ultrasonic Melt Treatment of Direct-Chill (DC) Casting. *Ultrason. Sonochem.* **2019**, *54*, 171–182. [CrossRef]
33. Ning, J.J.; Xiao, Q.L.; Dong, F.; Huang, M.Z. Effect of Ultrasonic Field on Undercooling Nucleation of TiC Particles in AlTiC Refiner. *Mater. Sci. Eng. Powder Metall.* **2015**, *20*, 19–25.
34. Ding, W.W.; Xia, T.D.; Zhao, W.J.; Hou, Y.F. Refining Performances of TiC and TiAl$_3$ Phases in Master Alloys on Pure Aluminum. *Chin. J. Nonferrous Met.* **2009**, *19*, 1025–1031.
35. Han, Y.; Ke, L.; Wang, J.; Da, S.; Sun, B. Influence of High-Intensity Ultrasound on Grain Refining Performance of Al–5Ti–1B Master Alloy on Aluminum. *Mater. Sci. Eng. A* **2005**, *405*, 306–312. [CrossRef]

© 2019 by the authors. Licensee MDPI, Basel, Switzerland. This article is an open access article distributed under the terms and conditions of the Creative Commons Attribution (CC BY) license (http://creativecommons.org/licenses/by/4.0/).

Article

Role of Acoustic Streaming in Formation of Unsteady Flow in Billet Sump during Ultrasonic DC Casting of Aluminum Alloys

Sergey Komarov * and Takuya Yamamoto

Graduate School of Environmental Studies, Tohoku University, Miyagi 980-8579, Japan; takuya.yamamoto.e6@tohoku.ac.jp
* Correspondence: komarov@material.tohoku.ac.jp; Tel.: +81-22-795-7301

Received: 1 October 2019; Accepted: 23 October 2019; Published: 28 October 2019

Abstract: The present work investigated melt flow pattern and temperature distribution in the sump of aluminum billets produced in a hot-top equipped direct chilling (DC) caster conventionally and with ultrasonic irradiation. The main emphasis was placed on clarifying the effects of acoustic streaming and hot-top unit type. Acoustic streaming characteristics were investigated first by using the earlier developed numerical model and water model experiments. Then, the acoustic streaming model was applied to develop a numerical code capable of simulating unsteady flow phenomena in the sump during the DC casting process. The results revealed that the introduction of ultrasonic vibrations into the melt in the hot-top unit had little or no effect on the temperature distribution and sump profile, but had a considerable effect on the melt flow pattern in the sump. Our results showed that ultrasound irradiation makes the flow velocity faster and produces a lot of relatively small eddies in the sump bulk and near the mushy zone. The latter causes frequently repeated thinning of the mushy zone layer. The numerical predictions were verified against measurements performed on a pilot DC caster producing 203 mm billets of Al-17%Si alloy. The verification revealed approximately the same sump depth and shape as those in the numerical simulations, and confirms the frequent and large fluctuations of the melt temperature during ultrasound irradiation. However, the measured temperature distribution in the sump significantly differed from that predicted numerically. This suggests that the present mathematical model should be further improved, particularly in terms of more accurate descriptions of boundary conditions and mushy zone characteristics.

Keywords: ultrasonic DC casting; acoustic streaming; aluminum alloy; mathematical model; unsteady flow phenomena; sump evolution; mushy zone; experimental verification

---

1. Introduction

Ultrasonic casting is receiving increasingly more attention, because it provides relatively simple and effective way to control solidification structure of various alloys. There has been an especially great progress in the development of ultrasonic casting technology for light metals, particularly aluminum alloys. The main achievements in this area have has been summarized in a number of books [1,2] and review papers [3,4]. It is well known from the relevant literature that most of the ultrasound-assisted technologies exploit acoustic cavitation, the phenomenon of nucleation, growth, and violent implosion of tiny bubbles. There is a large volume of literature devoted to acoustic cavitation, especially in fields such as sonochemistry, wastewater treatment, biotechnology, and mineral engineering [4–8].

Stable cavitation starts when the sound pressure exceeds a threshold value. This value is dependent on such factors as type of liquid, content of gas, and concentration of solid impurities in liquid. For instance, in tap water and commercially pure molten aluminum, this threshold was found to be attained when the peak-to-peak amplitudes of sonotrode vibrations becomes higher than 4–5 μm and 10–11 μm, respectively, at a frequency of 20 kHz [9]. Generally, such a low vibration amplitude can be produced using many types of commercially available ultrasonic equipment. However, when the treatment process needs to be scaled up to meet industrial requirements, a number of serious challenges arise. The first, and perhaps most difficult challenge, is that the size and shape of sonotrode, which is one of the main parts of powerful ultrasonic installations, cannot be arbitrarily changed. It is commonly known that acoustic cavitation occurs over a limited space near the ultrasonic radiating surface. However, a simple enlargement of the sonotrode area presents a considerable challenge, because the sonotrode design must meet resonance conditions imposing strict constraints on its size and shape. These constraints are especially severe for ultrasonic-assisted casting, because in this case the sonotrode material must resist well high-temperature cavitation erosion, thermal shock, and high temperature oxidation.

Another problem is occurrence of acoustic streaming, which is inevitably generated when ultrasound is irradiated in a liquid, especially at a commonly used frequency of 20 kHz. Acoustic streaming is defined as a steady fluid motion caused by the attenuation of ultrasonic waves in fluids. Although there are three types of acoustic streaming depending on the streaming scale, this study focused only on the large-scale streaming, because this type of streaming has the greatest effect on the liquid flow near the sonotrode. The main problem resulting from the acoustic streaming is that liquid flow near the sonotrode becomes hardly controllable. Particularly, in continuous or semi-continuous ultrasonic casting, i.e., when molten metal can pass through the cavitation zone only once, the flow control near the sonotrode surface is one of the key factors influencing the process efficiency. A typical example is the direct chilling (DC) casting process in the aluminum industry. In this case, the best flow condition would be to ensure that all or at least most of the amount of the molten metal can be passed through the zone of intense cavitation located near the sonotrode working tip. This condition is especially important when ultrasonic vibrations are applied to the melt aiming at dispersing particles of grain refiners.

Another feature related to the large-scale acoustic streaming may be a forced convection of liquid metal which may affect the heat transfer during the metal solidification process. This may cause problems when ultrasound is irradiated into a pool of solidifying melt, for example in the DC casting process. In this case, the heat convection may result in a temperature drop in the molten metal causing a deterioration of its solidification structure. This problem has been briefly discussed in our earlier investigation on the ultrasonic DC casting of Al-Si hypereutectic alloys [10]. These two issues, namely the effects of acoustic streaming on the passage of melt through the cavitation zone and on heat convection, suggest that the acoustic streaming should be properly controlled in ultrasonic-assisted casting processes. However, our understanding of the underlying relationships influencing the above phenomena still remains unclear.

It should be noted that a lot of results on acoustic streaming and its related phenomena have been published recently in sonochemical literature. Although water or aqueous solutions were used as a liquid in these studies, it is a reasonable assumption that the features of acoustic streaming and mechanisms of its generation in water and molten aluminum are very similar. At least, the similarity of cavitation in water and aluminum was suggested in the earlier study [11]. Some features and underlying mechanisms of acoustic streaming generation have been investigated also in our earlier paper [12] both experimentally and numerically. In this paper, based on the well documented fact that acoustic cavitation and acoustic streaming are interrelated phenomena, a mathematical model of acoustic streaming has been proposed and experimentally verified by using water model and PIV (Particle Image Velocimetry) technique. Moreover, this paper contains a brief review of recent results of the above mentioned studies dealing with the sonochemical technology. Later on, the proposed model was used to numerically predict the size of cavitation zone and velocity of acoustic streaming in molten aluminum [13]. The results reveal that due to large difference in physical and acoustic properties, in aluminum melt attenuation of sound waves is larger and velocity of acoustic streaming is smaller than those in water if the sound pressure amplitude at the sonotrode tip is the same.

Nevertheless, studies on the acoustic streaming in molten aluminum and its effects on the solidification and structure of aluminum alloys are still scarce. The mere fact that acoustic streaming affects the solidification structure of aluminum alloys has long been known. Results of recent and earlier studies concerning the acoustic streaming in molten metals have been summarized in the above cited books [1,2]. However, experimental difficulties in measuring flow velocity under very high temperatures have limited earlier attempts to investigate the acoustic streaming in molten metals. Although experimental investigations of acoustic streaming in molten metals still predict a big challenge, current progress in supercomputer's capabilities and CFD techniques has made possible very accurate and realistic numerical simulation of multiphase fluid flows, heat, and mass transfer phenomena including the acoustic flows in bubbly liquids and molten metals as well as solidification phenomena. Thus, numerical simulation in combination with cold physical modeling, for example using water models, provides a very powerful tool for investigating the ultrasonic-assisted casting processes.

As mentioned above, acoustic streaming in molten metals and its underlying mechanisms have much in common with those in other liquids including water. Therefore, the results of earlier studies [14–16] on numerical simulation of acoustic streaming in water systems are of considerable importance for ultrasonic casting technology. Additionally, the recent works of Eskin et al. [17–19] have made a great contribution in understanding the mechanism and features of acoustic streaming in molten aluminum. For instance, in one of the studies [19], they developed a mathematical model to simulate an acoustic flow in the sump of a DC caster, and on this basis predicted that the acoustic flow causes an increase in the cooling rate at the solidification front, which eventually results in refinement of solidification structure of A6060 alloy. It is noteworthy that this significant effect of microstructure refinement was achieved despite a relatively small and low-amplitude sonotrode used in this study.

The goal of the present study is to investigate the effects of acoustic streaming on the distribution of temperature and flow pattern in the sump of DC cast billets when different types of hot-top unit are used. As we have reported in our previous study [10], the effects of ultrasonic treatment of molten aluminum are very dependent on what fraction of melts can be passed through the cavitation zone, which is the more intensive in the immediate vicinity of sonotrode tip. This is especially important when the ultrasound vibrations are applied to disperse and activate refiner particles immediately before solidification or during the casting of metal. The present study includes three parts. In the first part, acoustic streaming was investigated experimentally using water models. The second part was devoted to development of mathematical model to simulate ultrasonic-assisted DC casting with emphasis on flow and heat transfer in the sump. In the third part, the model proposed was verified by using the results obtained in a pilot-scale DC caster.

2. Experimental Methods and Procedure

2.1. Ultrasonic Equipment

The ultrasonic system used in this study comprises a commercially available ultrasonic generator (DG2000, Telsonic, Zurich, Switzerland) equipped with a piezoceramic converter and a dumbbell-shaped sonotrode made of Si_3N_4-based ceramics. Details on the sonotrode specification and characteristics can be found in our previous paper [20]. The working frequency of the ultrasonic system was automatically tuned to a frequency in the range of 18.9–21.2 kHz. The vibration amplitude of sonotrode tip was increased from 25 to 70 μm (p-p) in proportion to increasing the generator electric output power from 50 to 100%. The maximum power of generator was 2 kW. The amplitude was measured in air by using a Laser Displacement Sensor (LK-G5000, Keyence, Osaka, Japan).

2.2. PIV Measurements

Particle Image Velocimetry (PIV) system was applied to measure the acoustic streaming velocity. Figure 1 is a schematic drawing of the experimental setup. A rectangular glass vessel (W 30 × L 30 × D 40 cm^3) was filled with water. Fluorescent particles (FLUOSTAR 0459, Kanomax Japan Inc., Shimizu, Japan) in an amount of 0.1 g were added to the water bath as a tracer. The average diameter of particles was 15 μm, and density was very close to that of water. The sonotrode tip was immersed vertically into the water bath to introduce ultrasonic vibrations followed by generation of cavitation and acoustic streaming. A water-cooled Ar lighting source (Optronics, Co., Ltd, Tokyo, Japan) was exploited to produce a visual laser sheet through the water bath in the immediate vicinity of the sonotrode tip, as shown in Figure 1. A high-speed camera (Photron, Tokyo, Japan) was used to record the tracer particle movement at a frame rate of 500 fps and a shutter time of 1/1000 second. When the fluorescent particles are illuminated by the Ar green light laser (wavelength 550 nm), they emit bright and well discernible light at a wavelength of 580 nm. This technique allowed us to completely eliminate any influence of cavitation bubbles on the PIV measurements. Then, the images taken were processed by a flow analysis software (Library Co., Ltd, Tokyo, Japan) to evaluate the acoustic streaming vector field. Figure 2 show typical results on the distribution of acoustic streaming velocity measured at a vibration amplitude of 50 μm (p-p). A comparison of these results with these published in our earlier paper [12] reveals that the use of fluorescent particles makes it possible to obtain much clearer picture of acoustic streaming vectors even in the immediate vicinity of the sonotrode tip.

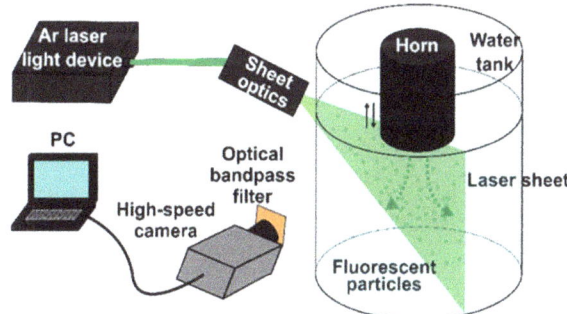

Figure 1. Experimental setup for Particle Image Velocimetry (PIV) measurements.

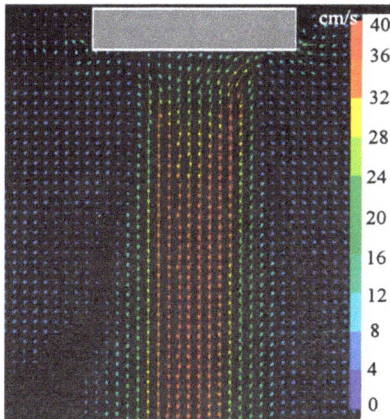

Figure 2. Typical results of PIV measurements.

2.3. Pilot DC Caster Tests

Pilot vertical DC casting facility was utilized in the present investigation using an Al-17%Si model alloy. The alloy was melted in an electrical resistance furnace and commonly used Al-Cu-P refiner was added to the melt. After degassing with a rotary argon degasser unit, the melt was poured through a launder into the hot-top equipped DC caster mold to produce billets of 203 mm in diameter and about 2 m in length by using conventional or ultrasonic-assisted casting techniques. In the latter case, the following two methods of ultrasonic treatment (hereinafter referred as to "UST") were examined. 1–UST inside hot-top, (UST HT) 2–UST inside the hot-top equipped by a bottom plate with a hole drilled at the plate center to let the melt flow in the mold at the center part of billet. This casting method will be referred as to UST in "flow-restricting hot-top" or UST FRHT. The DC caster of the latter type is schematically depicted in Figure 3. The sonotrode tip was preheated up to the melt temperature before its immersion into the melt. More details about these casting methods and the hot-top design can be found in our earlier study [10].

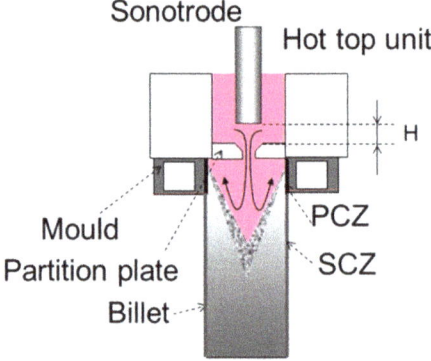

Figure 3. A schematic representation of flow-restricting hot-top for ultrasonic treatment.

In order to verify the results of numerical simulation, the profile of sump and distribution of the melt temperature in sump were measured in a number of experiments. In the profile measurements, a certain amount of Al-30%Cu master alloy was premelted in a separate furnace and hold at the temperature of Al-Si melt used for casting. In the final stage of the casting, the Al-Si melt supply was stopped and replaced with the Al-Cu melt flowing into the caster mold over a short period of time. Then, the produced billet was cut along the axial direction, and the cut surface was polished and etched to reveal the sump profile. The temperature distribution was measured in the vertical direction at different radii from the billet center axis according to the following procedure. The tip of a K-type thermocouple was attached to the one end of a thin metal wire, while the other end of the wire was fixed at the bottom block of DC caster before starting the casting. The wire was stretched vertically in such a way that the thermocouple tip was positioned initially above the hot-top unit. Hence, as the billet moves down during the casting, the thermocouple tip passed through the molten metal in the hot-top, sump, and mushy zone that made it possible to measure the temperature distribution in the vertical direction of billet.

3. Mathematical Model

The main aim of the present model was to predict flow pattern and temperature distribution in the sump of DC caster. In the case of conventional DC casting, the model includes the balance equations of mass, momentum, and heat transfer for fully liquid and mushy regions, and heat transfer for solid phase region. In the case of ultrasonic DC casting, a model describing cavitation-driven acoustic streaming in liquid was added to the system of balance equations.

3.1. Acoustic Streaming

Acoustic cavitation and acoustic streaming are interrelated phenomena, and therefore, they should be modeled within a single conceptual framework. In the present study, we used the model that was reported in our earlier study [12]. Therefore, only a brief explanation will be given here. The main simplifications and assumptions used for the simulations are as follows:

(1) All calculations were performed at a constant size of cavitation bubble, which can be interpreted as a size averaged over the ultrasound wave cycle.
(2) Cavitation bubbles were assumed to be filled with air or hydrogen when ultrasound was irradiated in water or molten aluminum, respectively. Physical properties of gases and liquids were set to be temperature independent.
(3) As the bubble volume ratio is very low, the two-phase liquid-bubble flow was treated as a single-phase flow.
(4) The fluids were assumed to be incompressible and Newtonian
(5) Liquid flow, cavitation zone, and heat transfer were assumed to be axially symmetric.
(6) Thus, the present model is incapable of simulating the cavitation phenomena themselves, but it can predict and model several important phenomena related to the acoustic cavitation. These phenomena include the sound pressure distribution, cavitation zone formation and acoustic streaming generation.

Propagation of ultrasound wave in a bubbly liquid can be described by the linearized Helmholtz equation as follows [15]:

$$\nabla^2 P + k_m^2 P = 0 \tag{1}$$

where P is the sound pressure amplitude and k_m is the complex wave number defined as:

$$k_m^2 = \frac{\omega^2}{c^2}\left(1 + \frac{4\pi c^2 N R_0}{\omega_0^2 - \omega^2 + 2ib\omega}\right) \tag{2}$$

where i is an imaginary unit. This equation reveals that the wave number is dependent on the wave angular frequency, ω, resonant frequency of bubble, ω_0, sound velocity in liquid, c, radius of undisturbed bubble, R_0, damping factor, b, and bubble number density, N. The bubble resonant frequency ω_0 can be determined from Equation (3):

$$\omega_0^2 = \frac{p_0}{\rho R_0^2}\left(Re\Phi - \frac{2\sigma}{R_0 p_0}\right) \tag{3}$$

where ρ is the liquid density, p_0 is the ambient pressure, σ is the surface tension, Re is the real part of Φ and Φ is the complex dimensionless parameter, which is the following function of the specific heat ratio of gas, γ, and a dimensionless parameter, χ

$$\Phi = \frac{3\gamma}{1 - 3(\gamma-1)i\chi\left[(i/\chi)^{1/2}coth(i/\chi)^{1/2} - 1\right]} \tag{4}$$

The latter is given as:

$$\chi = D/\omega R_0^2 \tag{5}$$

where D is the thermal diffusivity of gas. The damping factor b can be defined as:

$$b = 2\mu/\rho R_0^2 + p_0 Im\Phi/2\rho\omega R_0^2 + \omega^2 R_0/2c \tag{6}$$

where μ is the fluid dynamic viscosity.

The bubble number density, N is the following function of bubble radius and volume fraction of bubbles, β.

$$N = \frac{3\beta}{4R_0^3 \pi} \tag{7}$$

The volume fraction was considered to be proportional to the absolute value of the sound pressure magnitude, $|P|$.

$$\beta = C|P| \tag{8}$$

where the constant C was set to 2×10^{-9}. Both the relation (8) and the value of constant C were suggested by Jamshidi [15]. A constant value of $R_0 = 200$ µm was used in the present study.

Thus, in the present model, number density and volume fraction of cavitation bubbles are fully governed by the amplitude of sound pressure, frequency of ultrasound waves as well as bulk and surface properties of fluid. Obviously, both the number density and volume fraction of cavitation bubbles reach their maximum values in the vicinity of sonotrode tip where the sound pressure amplitude is the highest. Ultrasound waves are incident on the surface of cavitation bubbles forcing them to move away from the tip along the ultrasonic wave propagation. These moving bubbles entrain the surrounding liquid into a stream motion, which is commonly termed acoustic streaming. It is noteworthy that this hypothesis has often been used to explain the origin of acoustic streaming [21,22]. In a real situation, when cavitation bubbles are oscillated, the interaction between these oscillating bubbles and ultrasound waves is rather complicated, and force acting on the bubbles in this case is called the primary Bjerknes force. However, as in the present model the size of bubbles is kept constant, expression for the force may be significantly simplified. If the bubble is small, so that the change of the spatial gradient of sound pressure ∇P within the bubble volume is negligible, the force, F acting on the bubble can be expressed in terms of ∇P and the bubble volume, V [23]:

$$F = -\langle V \nabla P \rangle \tag{9}$$

where the bracket indicates the time averaging. As has been discussed in our earlier work [12], in the case of enough strong acoustic field under the sonotrode tip, this equation can be further simplified and rearranged to give the final form usable in the present simulation:

$$F = -\beta \nabla |P| \tag{10}$$

where β is the above-mentioned bubble volume fraction.

This equation was incorporated into the Navier-Stokes equation as an external volumetric force to simulate the acoustic streaming. The absolute value of sound pressure magnitude, $|P|$ was calculated using the above set of Equations (1)–(8). The following boundary conditions were used.

(1) Liquid flow velocity: no-slip and slip conditions for liquid flow at the solid walls and free surface, respectively.
(2) Sound pressure: full reflection condition at the solid surfaces.

3.2. Solidification Model

As has been mentioned above, the mathematical model, proposed in the present study, attempts to predict unsteady flow and heat transfer phenomena during the DC casting process. In order to overcome the computation time inconvenience, a number of simplifications were made. Particularly, temperature and composition dependences of the physical properties of melt and solidified phases were ignored. Instead, average values for Al-17%Si alloy were applied in the present model. Moreover, the mushy zone was assumed to have the same porous morphology, without subdividing it into slurry and solid network parts.

In the present study, the solidification model uses an enthalpy-porosity method of error-function model without considering shrinkage phenomena [24]. Additionally, the unsteady phenomena, including the movement of bottom block and solidified zone, are taken into account. The following porosity drag force is introduced into the Navier-Stokes equation as an external force:

$$F = -C' \frac{(1-\alpha)^2}{\alpha^3 + b}(u - u_b) \tag{11}$$

where C' is the drag constant, α is the solid fraction, b is the numerical stabilizing constant, u is the velocity, and u_b is the bottom block velocity. In the present model, the moving coordinate was used only for the solidified region. The solid fraction α is described as:

$$\alpha = \frac{1}{2}\text{erf}\left(\frac{4(T-T_m)}{T_l - T_s}\right) + \frac{1}{2} \tag{12}$$

where erf indicates the error function, T is the temperature, T_m is the mean temperature between the solidus, T_s, and liquidus, T_l, temperatures.

The following latent heat term is also introduced in the energy equation as:

$$S = -\rho L \frac{4\exp\left(-\left(\frac{4(T-T_m)}{T_l - T_s}\right)^2\right)}{(T_l - T_s)\sqrt{\pi}} \cdot \left(\frac{\partial T}{\partial t} + u \cdot \nabla T\right) \tag{13}$$

where S is the latent source term due to phase change and L is the latent heat.

Physical properties of molten metal necessary for the simulation are presented in Table 1. It is to be noted that no turbulent model was included in the present simulation. Instead, direct numerical simulation was applied over the fluid computational domain with very fine mesh. The following heat transfer coefficients were set as boundary conditions at the mold inner surface (primary cooling zone, PCZ) and water-cooled part of biller surface (secondary cooling zone, SCZ). Primary cooling zone, h_m = 10 kW/m²·K, secondary cooling zone, h_b = 10 kW/m²·K. The zone locations are indicated in Figure 3 as PCZ and SCZ, respectively.

4. Results

4.1. Verification of Acoustic Streaming Model

Figure 4 presents typical numerical results on acoustic streaming velocity in water (Figure 4a) and molten aluminum (Figure 4b). The vibration amplitude of sonotrode tip was set to 50 μm (p-p), which is the same value as in the above-mentioned PIV experiment. Predicted velocity vectors for water indicate a reasonable agreement with the experimental observations (Figure 2), although the acoustic flow obtained by the PIV measurement is slightly diffusive compared with the predicted one. Pattern of acoustic streaming in molten aluminum is very similar to that in water; however, the streaming velocity is significantly slower. The reason for this has been discussed in our earlier paper [12].

4.2. Temperature Distribution and Sump Characteristics

Next, using the above mathematical models, unsteady numerical simulations were conducted to predict evolution of the sump depth and profile over time as well as time-dependent distributions of temperature and velocity in the sump during the casting process. The total time of simulation was set to 3 min with an interval of 1 s. Figure 5 shows typical temperature contours in the sump in the last 180th second of simulation for the above-mentioned three cases. These contours are snapshots of supplemental Videos S1–S3 showing the time evolution of sump and temperature. The diameter of billets was 203 mm and the height of the mold was 30 mm, as shown by a grey rectangle in Figure 5c.

The melt initial temperature and casting speed was set to 780 °C and 200 mm/min, respectively. Two black rectangles at the upper part of Figure 5 indicate the location of sonotrode tip. Two white curves at each temperature contour indicate the locations of the liquidus and solidus profiles. Therefore, the area between these two lines corresponds to the mushy zone where the solid and liquid phases coexist. The results reveal that the sump depth and profile are independent of casting method. In the case of UST in the mold (Figure 5b), the sump looks slightly deeper as compared to the other conditions. However, this difference is rather small, as indicated in Figure 6. This figure shows time variation of the sump depth for all three conditions. The sump depth was defined as the distance from the bottom edge of the mold to the deepest location at the liquidus profile. The data of Figure 6 suggests that the sump depth increases rapidly in the beginning of casting, but then the increase gradually slows down and the depth becomes almost constant after 3 min of casting. Moreover, in Figure 5, the mushy zone formed during the conventional casting (Figure 5a) looks smaller than those formed under application of ultrasound oscillations. However, time variations of dimensionless volume of mushy zone, α presented in Figure 7 clearly reveal that this volume varies with time approximately in the same way for all casting methods examined in this study. The values of α were non-dimensionalized relative to volume of melt flowing into the sump from the mold. The only difference is that fluctuations of α values occurs more frequently under the UST conditions compared to the case of conventional casting. The reason for that will be discussed below.

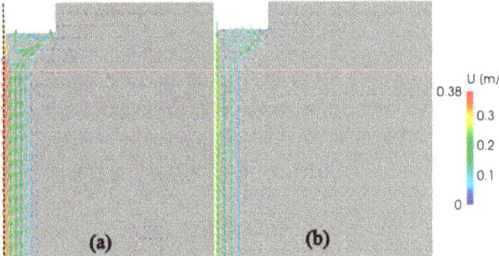

Figure 4. Predicted velocities of acoustic streaming in water (**a**) and molten aluminum (**b**).

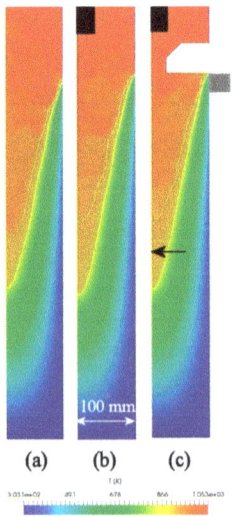

Figure 5. Distribution of temperature in sump. (**a**) Conventional casting, (**b**) UST HT casting, (**c**) UST FRHT casting.

Table 1. Model parameters used for the DC casting simulation.

Parameter	Unit	Quantity
Dynamic viscosity	kg·m^{-1}·s^{-1}	5.472×10^{-7}
Density	kg·m^{-3}	2330
Coefficient of thermal expansion	-	2.1×10^{-5}
Heat capacity	kJ·kg^{-1}·K^{-1}	1.282
Thermal conductivity	W·kg^{-1}·K^{-1}	80
Heat of fusion	kJ·kg^{-1}	450
Prandtl number	-	0.0166
Blake threshold	Pa	1.0×10^5
Liquidus temperature	K	920.15
Solidus temperature	K	850.15
Ultrasound frequency	kHz	20
Sound speed	m·s^{-1}	4650

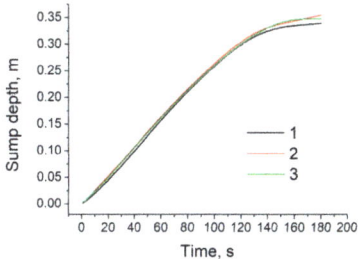

Figure 6. Time variation of sump depth. 1: conventional casting; 2: UST HT castingmaterials-625822 3: UST FRHT casting.

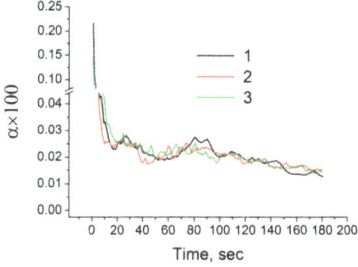

Figure 7. Time variation of dimensionless volume of mushy zone. 1: conventional casting; 2: UST HT casting; 3: UST FRHT casting.

Figure 8 presents the sump depth and profile measured experimentally in the case of UST FRHT casting. The casting conditions were the same as mentioned above. As it was difficult to identify the location of the mold from this picture, the sump depth can be measured only roughly. Under the given conditions, the sump depth is 300 mm approximately. It is to be noted that this value is much greater than those reported in the literature for close sizes of billet and the similar casting speed [25]. This is because in our experiments, a different alloy and much higher temperature of casting was used. However, as can be seen from Figure 8, at the lower part of the sump the mushy zone loses its continuity, suggesting that this is a partly or fully coherent solid part of the mushy zone. Therefore, the depth of sump may be shorter than 300 mm. The black horizontal arrow in Figure 5c shows the point corresponding to the 300 mm depth. It is clearly seen that this point is located inside the predicted mushy zone. This finding suggests a good agreement and predictability for our model. As the mushy zone was interpreted as being a porous solid phase, the predictions of sump depth give underestimated values in the present model. In the actual casting process, as a part of mushy zone is in form of flowable slurry, the actual sump should be deeper.

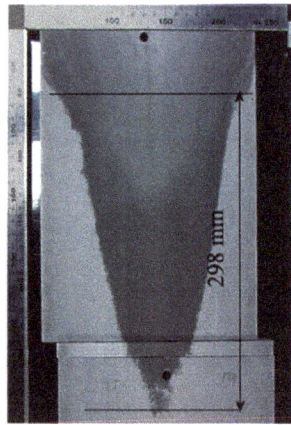

Figure 8. Profile of the sump in a billet produced by UST FRHT casting.

4.3. Distribution of Velocity in the Sump

The results of numerical simulation revealed that the velocity characteristics are considerably changed with the application of ultrasound vibrations. Figure 9 presents two typical sets of velocity vectors over the sump computed under conditions of conventional casting (a,d), UST HT casting (b,e), and UST FRHT casting (c,f) in the 65th (a,c) and 142nd (d,f) seconds of casting. These images are snapshots of supplemental Videos S4–S6, where more details on the velocity vectors can be found. In the case of conventional casting, the magnitude of velocity vectors is considerably smaller than in the cases of ultrasonic assisted casting. However, what is more important is that the ultrasonic vibrations make the melt flow inside the sump more turbulent. This can be readily seen from Figure 9b,c,d,f; a lot of eddies of small sizes are formed in the sump bulk and near the solidification front including the area close to the mold surface. The difference in the flow pattern between the conventional and UST castings is especially significant in the upper half of the sump. The eddy formation results in thinning of the mushy zone layer, as can also be seen from this figure. This phenomenon can cause the cooling rate to become higher that can contribute to the refinement of microstructure. The simulation results suggest that the eddy formation can occur also during the conventional casting as shown in Figure 9a. However, the eddy size in this case is much larger as compared to the UST casting cases. Another important finding is that the eddy formation becomes weaker with the distance from the sonotrode. As can be seen from Figure 9e,f, no eddies are produced in the lower part of the sump.

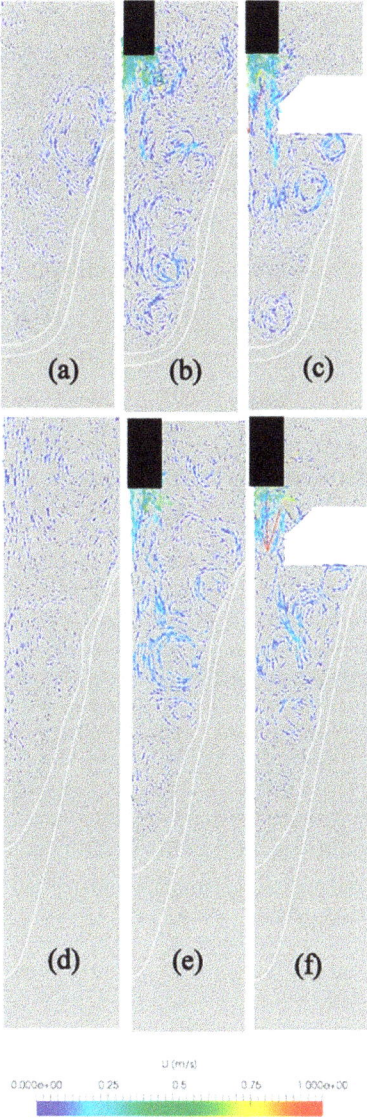

Figure 9. Predicted velocities in the sump during its evolution. (**a,d**) Conventional casting, (**b,e**) UST HT casting, (**c,f**) UST FRHT casting.

5. Discussion

The results of numerical simulations revealed that introduction of ultrasonic vibrations into the melt, flowing through the hot-top unit, has little or no effect on the temperature distribution and sump profile, but has a considerable effect on the melt flow pattern in the sump. Below is a discussion regarding the obtained simulation results and an additional verification of them versus experimental observations.

The reason why there is almost no difference in the temperature distribution, volume of mushy zone, and profile of the sump between these three casting methods is assumed to be as follows. In the present simulation, all boundaries except for the primary cooling surface (mold inner surface) and secondary cooling surface (water cooled part of billet) were set to be adiabatic. However, in the actual DC casting process, a part of the heat can escape through the melt free surface and walls at the hot-top unit. Moreover, it is well known that irradiation of ultrasound in liquids including molten metals is accompanied by generation of heat, which can be added to the melt [1]. In the present study, the heat generation was ignored. The fact that heat was lost through the melt free surface and hot-top unit walls can be demonstrated by the following experimental observations.

Figure 10 shows variations of temperature along the vertical distance when the thermocouple tip passes through the molten bath above the partition plate, partition plate (Figure 3), melt in sump, mushy zone and finally enters the body of solidified billet. In this figure, vertical broken line in the left part indicates location of the top surface of partition plate, two vertical broken lines in the middle part correspond to location of mold of height H_M = 30 mm, and vertical broken line in the right part shows an approximate location of the sump depth. Two horizontal broken lines indicate the liquidus, T_L, and solidus, T_S, temperatures. The measurement details and hot-top arrangement have been described in the above section. The measurements were conducted using the flow-restricting hot-top without and with ultrasound application. In the first case, two thermocouples were set in such a way that the first one passed along the billet centerline, while the second one passed at a distance of half-radius (R/2) from the center line. In the case of UST casting, the temperature measurements were performed only at the half-radius distance because the presence of sonotrode made is impossible to conduct the measurements along the centerline. In the casting without UST, the thermocouples enter the melt at a temperature of 780 °C, which remains almost the same till the thermocouple tips enter the melt in the sump (centerline location) or the partition plate (half-radius location). In the latter case, the measurements show that temperature inside the partition plate is reduced by 70 °C. After entering the melt inside the sump, the thermocouple indications become significantly dependent on the locations. At the center line (line 1), the melt temperature is slowly decreased until the upper boundary of the mushy zone is reached. It should be pointed that in this area, the temperature shows fluctuations that is in accordance with the results of numerical simulations. After entering the mushy zone, the temperature is abruptly decreased down, and then remains constant and close to T_L. This part of mushy zone is assumed to be fully flowable and its length is estimated to be about 100 mm. Then, as the thermocouple tip enters deeper into the mushy zone, temperature starts to lower (this point is shown by arrow in Figure 10), and finally drops to T_S. Most likely, these parts correspond to slightly coherent semisolid and fully coherent solid areas within the mushy zone. If one defines the sump as a zone in which molten metal can still flow, the sump depth becomes as shown by the right vertical line in Figure 10. At the R/2 location (line 2), the temperature drops more rapidly, and therefore, reaches the solidus temperature at a much shorter distance from the melt free surface compared to that at the centerline. However, under application of ultrasonic vibrations, the variation of temperature with distance is drastically changed (line 3). Firstly, the temperature fluctuates to a much greater extent than that during the casting without UST. This finding is in good agreement with the numerical predictions suggesting the generation of small eddies, especially at areas close to the mushy zone. Secondly, the temperature of the melt above the partition plate is significantly decreased with the distance from the melt surface. The most likely reason of this temperature drop is acoustic streaming which occurs in the melt above the partition plate enhancing heat loss through the melt free surface and hot-top walls. As a result, the temperature inside the partition plate becomes much lower as compared to that without UST. Thirdly, the temperature in the sump near the mold surface is decreased down to the liquidus temperature T_L, but then significantly increased up to 720 °C and decreased again to T_L with the distance from the partition plate. Finally, the temperature abruptly drops from T_L to T_S within a distance which is much shorter than in the case without UST suggesting a significant thinning of the mushy zone at the R/2 location. Additionally, these findings suggest that the sump, formed under conditions of ultrasound application, should have substantially greater width and, probably, depth as compared to the case without UST. However, these findings do not support the results of numerical simulation on the sump profile and temperature variation. Thus, although more experimental observations are needed to clarify this discrepancy, the experimental measurements of the present study suggest that our acoustic streaming model needs further improvements, particularly in terms of more accurate descriptions of boundary conditions and mushy zone characteristics.

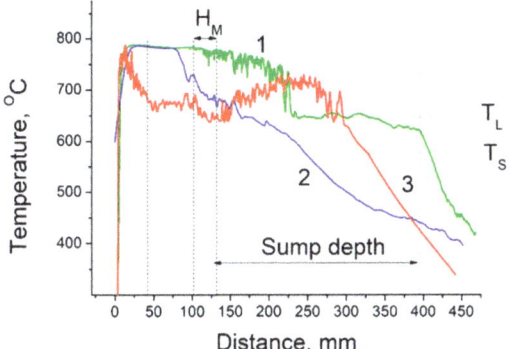

Figure 10. Variation of the melt temperature with distance from the melt surface. 1,2—FRHT without ultrasound irradiation, 3—FRHT with ultrasound irradiation.

It should be pointed out that the method of ultrasonic treatment of the melt in the flow-restricting hot-top unit has been developed in attempt to improve the efficiency of cavitation treatment of molten metal, particularly refiner particles in the melt, and to control the melt flow pattern in the sump. As reported in our earlier work [10], this method allows us to get two effects at the same time, namely the structural refinement and uniformity. In light of the results of the present numerical simulation, acoustic streaming, when occurs in the sump of molten aluminum, results less in the enhancement of convective flow than in the generation of turbulent eddies of relatively small sizes in the sump. It should be noted that this effect differs from that reported by Lebon et al. [19]. In their study, the tip of a sonotrode was positioned closer to the sump bottom, and therefore, the role of convection flow driven by the acoustic streaming was much greater than under conditions of our study. It is assumed that the generation of turbulent eddies in the sump is favorable from two points of view. First, the eddies can improve the dispersion of refiner particles and enhance their transport to the mushy zone. Second, as can be seen from Figure 9, small eddies are produced in the immediate vicinity of the mold surface, especially in the case of UST in the flow-restricting hot-top. This may be effective in controlling the billet surface defects, particularly the ripple mark. Our preliminary observations showed that the appearance of ripple marks on the billet surface are remarkable changed depending on the casting method and conditions. These issues are the subject of our ongoing research.

6. Conclusions

In the present work, melt flow pattern and temperature distribution in the sump of aluminum billets produced in a hot-top equipped DC caster conventionally and with ultrasonic irradiation were investigated, with the main emphasis being placed on the effects of acoustic streaming and hot-top unit type. As direct observations of flow phenomena in molten aluminum is impossible, the acoustic streaming characteristics were predicted by using the earlier developed numerical model verified against results of water model experiments. Then, the acoustic streaming model was applied to develop a simplified numerical model, which was capable, for the first time, of simulating unsteady phenomena in the sump during DC casting process. The results of numerical simulations were verified against measurements performed on an ultrasonic-assisted DC caster producing 203 mm billets of Al-17Si alloy at a casting speed of 200 mm/min. The results of the present study reveal the following.

(1) According to the numerical simulations, the evolution of the melt sump continues about 3 min irrespective of the ultrasound application and hot-top type. The final depth and the shape of the sump are also independent of the casting conditions examined.

(2) Irradiation of ultrasound into the melt at the hot-top unit results in more frequent fluctuations of the temperature and mushy zone volume around their mean values as compared to the conventional DC casting case, although the mean values themselves remain the same in both the cases.

(3) The pattern of melt flow in the sump is drastically changed with the ultrasound application. The flow velocity becomes faster in general and a lot of relatively small eddies are produced in the sump bulk and near the mushy zone. The latter causes frequently repeated thinning of the mushy zone layer.

(4) The experimental verification revealed approximately the same sump depth and shape as those predicted by the numerical simulations, and confirmed the frequent and large fluctuations of the melt temperature during ultrasound irradiation. However, the measurements of the temperature distribution in the sump showed a significant difference between the cases of castings without and with ultrasound irradiation. This suggests that the present mathematical model should be further improved, particularly in terms of more accurate descriptions of boundary conditions and mushy zone characteristics.

Supplementary Materials: The following are available online at http://www.mdpi.com/1996-1944/12/21/3532/s1, Video S1: time variation of sump profile and temperature for conventional casting; Video S2: time variation of sump profile and temperature for UST HT casting; Video S3: time variation of sump profile and temperature for UST FRHT casting; Video S4: time variation of sump profile and melt flow for conventional casting; Video S5: time variation of sump profile and melt flow for UST HT casting; Video S6: time variation of sump profile and melt flow for UST FRHT casting.

Author Contributions: S.K. conceptualized the manuscript wrote its first draft of this manuscript with substantial inputs from T.Y. T.Y developed the mathematical model and conducted the numerical simulation. All authors provided input, reviewed and finalized the manuscript.

Funding: This research received no external funding.

Acknowledgments: The authors would like to acknowledge the casting technology development group of Nippon Light Metal Co., Ltd for technical support and providing the results of pilot DC casting.

Conflicts of Interest: The authors declare no conflict of interest.

References

1. Eskin, G.I.; Eskin, D.G. *Ultrasonic Treatment of Light Alloy Melts*, 2nd ed.; CRC Press: London, UK, 2014; pp. 1–330.
2. Abramov, O.V. *High-Intensity Ultrasonics: Theory and Industrial Applications*; Gordon and Breach Science Publishers: Amsterdam, The Netherlands, 1998; pp. 1–685.
3. Komarov, S.V.; Kuwabara, M.; Abramov, O.V. High Power Ultrasonics in Pyrometallurgy: Current Status and Recent Development. *ISIJ Int.* **2005**, *45*, 1765–1782. [CrossRef]
4. Eskin, D.G. Ultrasonic processing of molten and solidifying aluminium alloys: Overview and outlook. *Mat. Sci. Tech.* **2017**, *6*, 636–645. [CrossRef]
5. Brennen, C.E. *Cavitation and Bubble Dynamics*; Oxford Press: Oxford, UK, 1995; pp. 1–294.
6. Wan, M.; Feng, Y.; Ter Haar, G. *Cavitation in Biomedicine: Principles and Techniques*; Springer: Heidelberg, Germany, 2015; pp. 1–503.
7. Bagal, M.V.; Gogate, P.R. Wastewater treatment using hybrid treatment schemes based on cavitation and Fenton chemistry: A review. *Ultrason. Sonochem.* **2014**, *21*, 1–14. [CrossRef] [PubMed]
8. Gogate, P.R.; Kabadi, A.M. A review of application of cavitation in biochemical engineering/biotechnology. *Biochem. Eng. J.* **2009**, *44*, 60–72. [CrossRef]
9. Komarov, S.V.; Oda, K.; Ishiwata, Y.; Dezhkunov, N. Characterization of Acoustic Cavitation in Water and Molten Aluminium Alloy. *Ultrason. Sonochem.* **2013**, *20*, 754–761. [CrossRef] [PubMed]

10. Komarov, S.V.; Ishiwata, Y.; Mikhailov, I. Industrial Application of Ultrasonic Vibrations to Improve the Structure of Al-Si Hypereutectic Alloys: Potential and Limitations. *Metall. Mater. Trans. A* **2015**, *46*, 2876–2883. [CrossRef]
11. Tzanakis, I.; Lebon, G.S.B.; Eskin, D.G.; Pericleous, K.A. Characterizing the cavitation development and acoustic spectrum in various liquids. *Ultrason. Sonochem.* **2017**, *34*, 651–662. [CrossRef] [PubMed]
12. Fang, Y.; Yamamoto, T.; Komarov, S.V. Cavitation and acoustic streaming generated by different sonotrode tips. *Ultrason. Sonochem.* **2018**, *48*, 79–87. [CrossRef] [PubMed]
13. Yamamoto, T.; Komarov, S. Investigation on Acoustic Streaming During Ultrasonic Irradiation in Aluminum Melts. In Proceedings of the TMS Light Metals, San Antonio, TX, USA, 10–14 March 2019; pp. 1527–1531.
14. Trujillo, F.J.; Knoerzer, K. A computational modeling approach of the jet-like acoustic streaming and heat generation induced by low frequency high power ultrasonic horn reactors. *Ultrason. Sonochem.* **2011**, *18*, 1263–1273. [CrossRef] [PubMed]
15. Jamshidi, R.; Pohl, B.; Peuker, U.A.; Brenner, G. Numerical investigation of sonochemical reactors considering the effect of inhomogeneous bubble clouds on ultrasonic wave propagation. *Chem. Eng. J.* **2012**, *189–190*, 364–375. [CrossRef]
16. Louisnard, O. A simple model of ultrasound propagation in a cavitating liquid. Part II: Primary Bjerknes force and bubble structures. *Ultrason. Sonochem.* **2012**, *19*, 66–76. [CrossRef] [PubMed]
17. Salloum-Abou-Jaoude, G.; Eskin, D.G.; Lebon, G.S.B.; Barbatti, C.; Jarry, P.; Jarrett, M. Altering the Microstructure Morphology by Ultrasound Melt Processing During 6XXX Aluminium DC-Casting. In Proceedings of the TMS Light Metals, San Antonio, TX, USA, 10–14 March 2019; pp. 1605–1610.
18. Lebon, G.S.B.; Tzanakisc, I.; Pericleous, K.; Eskin, D.G.; Grant, P.S. Ultrasonic liquid metal processing: The essential role of cavitation bubbles in controlling acoustic streaming. *Ultrason. Sonochem.* **2019**, *55*, 243–255. [CrossRef] [PubMed]
19. Lebon, G.S.B.; Salloum-Abou-Jaoude, G.; Eskin, D.G.; Tzanakisc, I.; Pericleous, K.; Jarry, P. Numerical modelling of acoustic streaming during the ultrasonic melt treatment of direct-chill (DC) casting. *Ultrason. Sonochem.* **2019**, *54*, 171–182. [CrossRef] [PubMed]
20. Komarov, S.; Yamamoto, T. Development and Application of Large-Sized Sonotrode Systems for Ultrasonic Treatment of Molten Aluminum Alloys. In Proceedings of the TMS Light Metals, San Antonio, TX, USA, 10–14 March 2019; pp. 1597–1604.
21. Servant, G.; Caltagirone, J.P.; Ge´rard, A.; Laborde, J.L.; Hita, A. Numerical simulation of cavitation bubble dynamics induced by ultrasound waves in a high frequency reactor. *Ultrason. Sonochem.* **2000**, *7*, 217–227. [CrossRef]
22. Louisnard, O. A viable method to predict acoustic streaming in presence of cavitation. *Ultrason. Sonochem.* **2017**, *35*, 518–524. [CrossRef] [PubMed]
23. Doinikov, A.A. Bjerknes forces and translational bubble dynamics. In *Bubble and Particle Dynamics in Acoustic Fields: Modern Trends and Applications*; Doinikov, A.A., Ed.; Research Signpost: Thiruvananthapuram, India, 2005; pp. 2–49.
24. Rösler, F.; Brüggemann, D. Shell-and-tube type latent heat thermal energy storage: Numerical analysis and comparison with experiments. *Heat Mass Trans.* **2011**, *47*, 1027–1033. [CrossRef]
25. Eskin, D.G.; Zuidema, J.; Savran, V.I.; Katgerman, L. Structure formation and macrosegregation under different process conditions during DC casting. *Mater. Sci. Eng. A* **2004**, *384*, 232–244. [CrossRef]

© 2019 by the authors. Licensee MDPI, Basel, Switzerland. This article is an open access article distributed under the terms and conditions of the Creative Commons Attribution (CC BY) license (http://creativecommons.org/licenses/by/4.0/).

Article

Numerical Modelling of the Ultrasonic Treatment of Aluminium Melts: An Overview of Recent Advances

Bruno Lebon [1,*], Iakovos Tzanakis [2], Koulis Pericleous [3] and Dmitry Eskin [1]

1. Brunel Centre for Advanced Solidification Technology, Brunel University London, Kingston Lane, Uxbridge UB8 3PH, UK; Dmitry.Eskin@brunel.ac.uk
2. Oxford Brookes University, Wheatley Campus, Oxford OX33 1HX, UK; itzanakis@brookes.ac.uk
3. Computational Science and Engineering Group, University of Greenwich, 30 Park Row, London SE10 9LS, UK; K.Pericleous@greenwich.ac.uk
* Correspondence: Bruno.Lebon@brunel.ac.uk

Received: 17 September 2019; Accepted: 3 October 2019; Published: 6 October 2019

Abstract: The prediction of the acoustic pressure field and associated streaming is of paramount importance to ultrasonic melt processing. Hence, the last decade has witnessed the emergence of various numerical models for predicting acoustic pressures and velocity fields in liquid metals subject to ultrasonic excitation at large amplitudes. This paper summarizes recent research, arguably the state of the art, and suggests best practice guidelines in acoustic cavitation modelling as applied to aluminium melts. We also present the remaining challenges that are to be addressed to pave the way for a reliable and complete working numerical package that can assist in scaling up this promising technology.

Keywords: numerical modelling; acoustic cavitation; aluminium; ultrasonic melt treatment; non-linear bubble dynamics; sonoprocessing

1. Introduction

Process design has the potential to provide a strategic competitive advantage with regards to customer appeal, product cost and innovation [1,2]. A key element of process innovation involves a fundamental understanding of how materials and process interactions determine manufacturing performance [2].

A continuous mode of production is often more desirable than batch production. Advantages of continuous operations include cheaper unit costs of production, energy savings and homogenization in the quality of the product. However, converting batch processes to continuous modes is not straight-forward [3]. In the past six years, the authors have been researching how to upscale the promising technology of ultrasonic melt processing by moving applications from batch mode to inline mode [4].

Ultrasonic melt processing (USP) is an effective method for degassing, filtration and grain refinement of light metal alloys on the industrial scale [5–7]. The beneficial effects of USP are attributed to acoustic cavitation—the violent pulsation and collapse of gas bubbles under the influence of a strong acoustic pressure field [8,9], and acoustic streaming—the fluid motion that results from the attenuation of the acoustic pressure wave as it propagates in the liquid [10]. While USP works well in batch degassing or grain refining of a single cast billet or ingot in direct-chill (DC) casting [11], it does not scale up very well for continuous processing, unless multiple ultrasound sources are used [4]. Current research is now focusing on upscaling this promising technology and achieving high efficiency when treating large melt volumes in continuous mode and with a minimum number of ultrasound sources.

To optimize USP, recirculation patterns and mass exchanges between the cavitation zone and the rest of the liquid bulk need to be adequately quantified. In addition, melt recirculation reduces temperature gradients and promotes a preferred equiaxed grain structure [12]. In DC casting, acoustic streaming improves chemical homogeneity and promotes grain refinement through deagglomeration of clusters, wetting of inclusions, dispersion of substrates and solid-phase fragmentation [7,13,14]. However, it is challenging to visualize acoustic streaming in liquid aluminium due to its opaqueness and high operating temperature. Therefore, studies of acoustic streaming must include modelling in conjunction with X-ray imaging [15–18] to predict the generation of cavitation bubbles, their transport and acoustic propagation in the presence of attenuation.

A recent review of acoustic pressure modelling presents the challenges facing acoustic cavitation modelling [19] and has been the starting point for the model that resulted from a joint collaboration between Brunel and Greenwich universities in the U.K. [4]. Since then, a novel 'advanced' model that incorporates acoustic streaming with cavitation dynamics has been made available [20], enabling more accurate predictions of processing simulations involving cavitation bubbles at an affordable computational cost. This recent progress is summarised in this overview giving readers a good starting point in ultrasonic process modelling.

2. Existing Models

2.1. Acoustic Cavitation Model

The theory is reproduced here as a comprehensive summary of acoustic cavitation treatment in modelling of ultrasonic melt processing. The starting point of acoustic cavitation modelling is the Caflisch equations [21]. Sound propagation in a liquid containing bubbles has been studied with a set of nonlinear equations postulated by van Wijngaarden in the 1960s [22], and then derived mathematically by Caflisch et al. [21] using Foldy's method [23]:

$$\frac{\partial p}{\partial t} + \rho_l c_l^2 \nabla \cdot \mathbf{u} = \rho_l c_l^2 \frac{\partial \phi}{\partial t}, \text{ and} \quad (1)$$

$$\rho_l \frac{\partial \mathbf{u}}{\partial t} + \nabla p = \mathbf{F}, \quad (2)$$

where p is the acoustic pressure, $\mathbf{u}$ are the velocities, ρ_l is the (pure) liquid density, c_l is the speed of sound in pure liquid, $\phi = VN = \frac{4}{3}\pi NR^3$ is the bubble phase fraction for a bubbly system with a bubble density of N, consisting of identical bubbles each of radius R and V denotes the volume of a single bubble. The bubble density is assumed to be given by a step function:

$$N = \begin{cases} N_0 \text{ if } |P| > P_B \\ 0 \text{ if } |P| \leq P_B \end{cases}, \quad (3)$$

where $P_B = 1 + \sqrt{\frac{4}{27}\frac{S^3}{1+S}}$ is the Blake threshold [24,25], with the dimensionless Laplace tension $S = \frac{2\sigma}{p_0 R_0}$, σ is the surface tension between the liquid and gas, p_0 is the atmospheric pressure, and R_0 is the equilibrium radius of the bubbles. The momentum source term F can be used to prescribe acoustic velocity sources, e.g., due to the vibrating horn.

The radius of a bubble is obtained by solving a second-order ordinary differential equation (ODE). For an accurate bubble dynamics representation at high forcing pressures, compressibility has to be taken into account. For liquid metals, bubble dynamics can be represented by the Keller–Miksis equation [26]:

$$\rho_l\left[\left(1-\frac{\dot{R}}{c_l}\right)R\ddot{R} + \frac{3}{2}\left(1-\frac{\dot{R}}{3c_l}\right)\dot{R}^2\right] = \left(1+\frac{\dot{R}}{c_l}+\frac{R}{c_l}\frac{d}{dt}\right)\left[p_g+p_v-\frac{2\sigma}{R}-\frac{4\mu_l\dot{R}}{R}-p_0\{1-A\sin(\omega t)\}\right], \quad (4)$$

where p_v is the vapour pressure of the gas in the bubble, μ_l is the dynamic viscosity of the pure liquid, A is the normalized pressure amplitude (relative to p_0) and ω is the angular frequency of the ultrasonic source. Taking into account the effect of heat transfer during bubble dynamics [27,28], the gas pressure p_g is evaluated by solving the following ODE:

$$\frac{dp_g}{dt} = \frac{3}{R}\left[(\gamma-1)k\frac{dT}{dr}\bigg|_{r=R} - \gamma p_g \dot{R}\right], \tag{5}$$

where k is the thermal conductivity of the bubble gas, T is the temperature inside the bubble and γ is the polytropic exponent. When assuming adiabatic bubble pulsation, the polytropic exponent is given by $\gamma = 1.4$. The temperature gradient at the bubble surface is approximated linearly following the method of Toegel et al. in water [29]:

$$\frac{dT}{dr}\bigg|_{r=R} = \frac{T-T_\infty}{\sqrt{(RD)/\{3(\gamma-1)\dot{R}\}}}, \tag{6}$$

where T_∞ is the liquid bulk temperature and D is the gas diffusivity. The bubble temperature is solved for by using another ODE expressing the first law of thermodynamics:

$$c_v \dot{T} = 4\pi R^2 k \frac{T-T_\infty}{l_{th}} - p_g \dot{V}, \tag{7}$$

where the thermal diffusion length $l_{th} = \min\left(\frac{R}{\pi}, \sqrt{\frac{RD}{R}}\right)$ and c_v is the specific heat capacity of the gas.

Solving the Caflisch equations coupled with the above ODEs is very computationally intensive. In general, the acoustic pressure p is required to compute the momentum source term that corresponds to acoustic streaming. To optimize the computational procedure and reduce its time cost, the following approximation is used in recent works.

$\Re(Pe^{i\omega t})$ denotes the harmonic part of the acoustic pressure p. A nonlinear extension [25,30] of the linear Helmholtz equation originally derived by Commander and Prosperetti [31] from the Caflisch equations approximately describes the complex amplitude P as:

$$\nabla^2 P + K^2 P = 0. \tag{8}$$

Commander and Prosperetti defined the complex wavenumber K using:

$$K^2 = \frac{\omega^2}{c_l^2}\left(1 + \frac{4\pi c_l^2 N R_0}{\omega_0^2 - \omega^2 + 2jb\omega}\right). \tag{9}$$

where ω_0 is the resonant frequency of the bubbles, j is the complex number satisfying $j^2 = -1$ and b is the damping factor defined elsewhere [31]. Louisnard [25] generalized this linear model while keeping realistic values for dissipation of energy in inertial cavitation, resulting in a nonlinear model. However, this model suffers from two deficiencies: (1) the real part of K^2 was taken to approach that of Commander and Prosperetti, and (2) the Helmholtz equation used by the model comes from the linear theory. To address these issues, the real and imaginary parts of the coefficient K^2 have been rigorously re-derived by Trujillo [30] as:

$$\Re(K^2) = \frac{\omega^2}{c_l^2} - \frac{\mathcal{A}}{|P|}, \text{ and} \tag{10}$$

$$\Im(K^2) = -\frac{\mathcal{B}}{|P|}, \tag{11}$$

where

$$\mathcal{A} = -\frac{\rho_l \omega^2}{\pi}\int_0^{2\pi}\frac{\partial^2 \phi}{\partial \tau^2}\cos\left(\tau + \frac{\pi}{2}\right)d\tau, \tag{12}$$

$$\mathcal{B} = \frac{\rho_l \omega^2}{\pi} \int_0^{2\pi} \frac{\partial^2 \phi}{\partial \tau^2} \sin\left(\tau + \frac{\pi}{2}\right) d\tau, \text{ or} \tag{13}$$

$$\mathcal{A} = -\frac{\rho_l \omega^2}{\pi} \int_0^{2\pi} \frac{\partial \phi}{\partial \tau} \sin\left(\tau + \frac{\pi}{2}\right) d\tau, \tag{14}$$

$$\mathcal{B} = -\frac{\rho_l \omega^2}{\pi} \int_0^{2\pi} \frac{\partial \phi}{\partial \tau} \cos\left(\tau + \frac{\pi}{2}\right) d\tau, \tag{15}$$

where the non-dimensional time τ is within one period, i.e., $[0, 2\pi]$. The boundary conditions for the nonlinear Helmholtz equation are generally defined as:

- $\nabla P \cdot n = 0$ for infinitely hard boundaries (such as crucible walls);
- $P = AP_0$ at the surface of the sonotrode;
- Setting $P = 0$ in the cell layer above the liquid level to approximate the π phase shift that occurs upon reflection from the free surface [32].

This nonlinear Helmholtz equation is rather simple to solve using the finite element method (FEM), thereby facilitating the numerical evaluation of the acoustic field in the presence of cavitation bubbles [19,20,25,30] in commercial packages such as COMSOL Multiphysics. The solution of this equation using the finite volume method (FVM) is trickier and requires special preconditioning [32]; however, the flow equations are simpler to solve in the FVM framework.

2.2. Macroscopic Flow Model

Acoustic streaming models in the literature generally follow the work of Eckart [33] by incorporating the streaming force f as:

$$f = -\nabla(\rho_l \overline{v \otimes v}), \tag{16}$$

where v is the acoustic velocity to the continuity and momentum conservation equations, leading to:

$$0 = \nabla \cdot (\rho_l u) + \nabla \cdot (\overline{\rho_v v}), \tag{17}$$

$$0 = f - \nabla p + \mu_l \nabla^2 u, \tag{18}$$

where ρ_v is the density variation that is caused by the primary acoustic field. However, Equation (18) is the momentum equation of a creeping flow driven by the acoustic streaming and is therefore applicable to Reynolds numbers much smaller than 1 [20]. Since acoustic cavitation processes would involve much larger Reynolds numbers, the streaming velocity should instead be calculated from a full steady-state Navier–Stokes equation [34]:

$$\nabla(\rho_l u \otimes u) = f - \nabla p + \mu_l \nabla^2 u. \tag{19}$$

However, the solution of Equation (19) is difficult since the streaming flow observed experimentally is turbulent [35], and resolving small-scale eddies with the Navier–Stokes equation is not trivial [20]. Instead of solving for streaming directly, the latest papers follow the approach of Louisnard [20] by computing the streaming force from the solution of the nonlinear Helmholtz Equation (8) and injecting the result into the momentum equation. This approach has been validated in recent works describing acoustic streaming [20,35–37] and has also been applied to DC casting [38].

3. Numerical Simulations of the Acoustic Field in Crucibles, Moulds and Launders

There is a dearth of contributions in the literature regarding the specific modelling of ultrasonic melt processing. This is not surprising since accurate measurements of acoustic pressures in aluminium have only recently been made available [39,40]. Some of the first modelling contributions are those of Nastac [41–43], who presented two approaches for modelling grain refinement of an A356 alloy. A similar approach is followed by other authors to model nanoparticle dispersion [44] and the distribution of acoustic pressure in a launder [45]. The first method consists of solving the Reynolds-Averaged Navier-Stokes (RANS) equations using a classical hydrodynamic cavitation model [46] that is implemented in commercial Computational Fluid Dynamics (CFD) packages. The essence of this method can be summarized as follows. The liquid–bubble mass transfer is governed by a bubble transport equation in the following form:

$$\frac{D(\rho_b \phi)}{Dt} = R_G - R_C, \tag{20}$$

where ρ_b is the bubble density and the source terms R_G and R_C account for bubble growth and collapse, respectively. These source terms are calculated using the growth of a single spherical bubble based on a bubble dynamics model (e.g., Rayleigh–Plesset [47], Keller–Miksis [26], Neppiras–Noltingk [48]). This equation is coupled with the flow conservation equations, together with a suitable closure for turbulence. However, as beautifully as it is presented by Louisnard [20], this model is restricted to bubbly liquids containing vapour bubbles only. With the vapour pressure of aluminium at the melting point being negligible [49], it is unlikely that aluminium vapour bubbles would form in the melt bulk [50] with gas, hydrogen-filled bubbles, forming instead. This therefore, prohibits the use of any hydrodynamic cavitation models for the inertial acoustic cavitation bubbles that are present in liquid aluminium treatment.

The second approach in Nastac's contribution is, however, more appropriate and is a precursor to the method highlighted in this overview. In this indirect method, the acoustic field is solved by using a linear Helmholtz equation, closed with the Neppiras–Noltingk model [48]. However, as argued in the previous section, this method suffers from various deficiencies: the linear Helmholtz model is inadequate in the presence of cavitation bubbles, since pressure propagation is nonlinear in this regime. The Neppiras–Noltingk model does not account for acoustic radiation, which is crucial at high forcing pressures. Other authors also used a linear acoustic propagation model to study the treatment of AlSi7Mg alloy melt in sand casting, even though pressures larger than 2 MPa have been predicted [51]. A linear model was also employed to compute the acoustic pressure field in a SCN-1 wt% camphor alloy, which is often used as a transparent analogue to aluminium melt [13,52].

Another attempt to obtain an accurate prediction of the acoustic pressure field was through the solution of the Caflisch equations (Equations (1) and (2)) [38]. In this approach, Lebon et al. directly computed the acoustic field using the nonlinear equations governing sound propagation in bubble liquids [21,53,54] and validated the model using experimental data from the literature as shown in Figure 1.

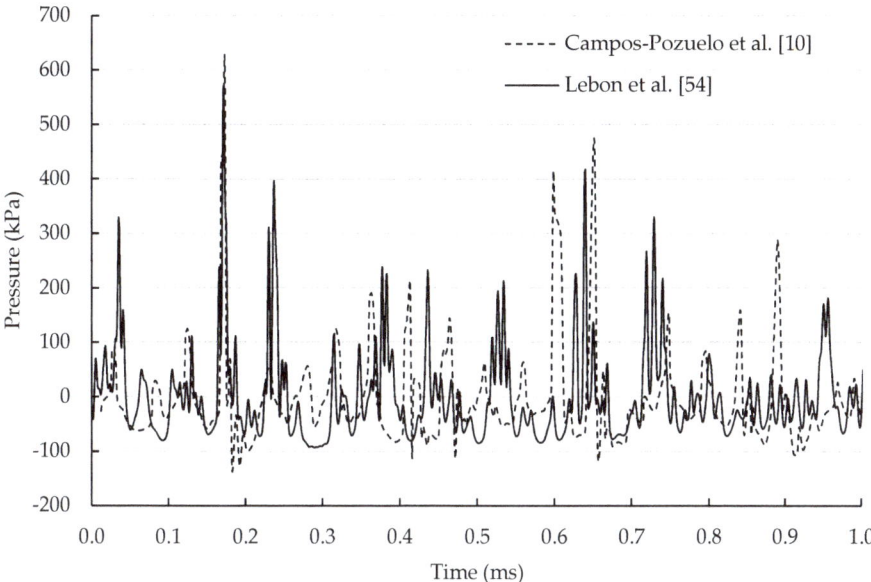

Figure 1. Validation of Caflisch approach to computing acoustic pressures [54] using experimental data from Campos-Pozuelo et al. [10].

Adequate pressures were predicted as compared with the measurements using a calibrated cavitometer [39,40] enabling the extension of the model to account for nanoparticle deagglomeration [55], fragmentation of dendrites [56], the erosion of thermally-sprayed splats [57] or contactless ultrasound due to Lorentz forces from an electromagnetic coil [58]. However, this method suffers from various drawbacks. A bubble dynamics ODE must be solved in each computational cell of the domain, with the use of an adaptive time-stepping scheme to stabilize the solution procedure for each computational cell. The method is also prone to numerical diffusion and requires the use of high-order discretisation schemes in space and time, and a special staggering scheme [59]. These issues prohibit the application of the model to complex 3D geometries due to the extreme computational requirements.

More recently, nonlinear models of pressure propagation have been used in the context of melt processing. Nonlinear models are required to more adequately capture the attenuation of pressure due to the presence of cavitating bubbles as shown in Figure 2. Huang et al. [60] used an improved nonlinear Helmholtz model [61] to predict the cavitation depth in sonication of an AlCu melt. Lebon et al. used Louisnard's model to compute acoustic pressures in water and aluminium vessels [35]. This model was adapted and improved to model acoustic pressures in DC casting [38]. The same approach has also been used by Yamamoto and Komarov [37]. The last two studies are conducted in conjunction with acoustic streaming and are discussed in the next section.

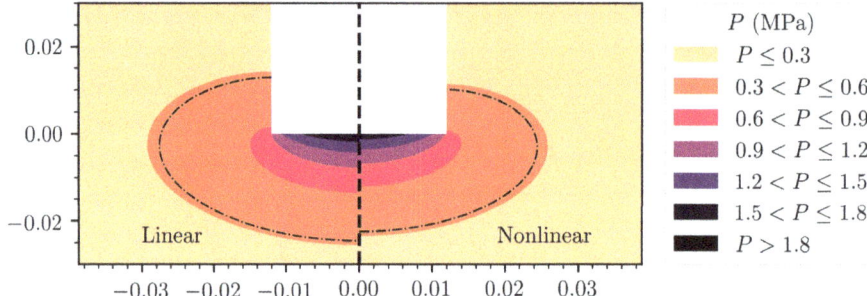

Figure 2. The difference in predicted acoustic pressures between linear and nonlinear models. The nonlinear model includes the attenuating effect of cavitating bubbles below the sonotrode. The dash-dotted line denotes the Blake threshold for hydrogen bubbles in aluminium, with $R_0 = 3$ µm.

4. Effect of Acoustic Streaming

The numerical study of acoustic streaming in melt processing is even sparser in the literature. A prediction of ultrasonic DC clad casting using the Navier–Stokes equations only has been recently published [62]. Acoustic streaming is implemented by a direct solution of the sonotrode motion in ANSYS Fluent and applying the transient effects in steady-state equations, although the implementation details of the process are not described. The model does not include the effect of cavitation bubbles, so acoustic shielding is not considered. As reported in [63], this method does not appear to capture the flow reversal observed at certain irradiation powers.

Other significant contributions, however, employ acoustic streaming models computing the acoustic streaming force as per Equation (16). Simulation of convective flow for an Al-2% Cu alloy has been performed by Wang et al. [12,64,65] using the Lighthill approach [34]. This approach predicts a fast velocity jet below the sonotrode, and comparison with a corresponding experiment reveals that the fast streaming flattens the temperature gradient and promotes an equiaxed grain structure. Another approach involved the use of the Ffowcs Williams and Hawkings (FW—H) equation generally used to compute the propagation of aerodynamic noise to model the acoustic pressure in Fluent [66,67]. Another commercial CFD software package, Flow3D, has been used to model acoustic streaming during the treatment of an A356 alloy melt without detailing the modelling procedure [68].

Inspired by Louisnard's nonlinear model coupled with acoustic streaming, Lebon et al. [35] validated an acoustic streaming model using results from a particle image velocimetry (PIV) experiment using a TSI system [63] as shown in Figure 3. While the model offers only qualitative agreement with the experiment, this progress is encouraging because:

1. Despite simulating the problem in two dimensions, the correct order of magnitude of acoustic streaming is recovered.
2. The net flow reversal below the sonotrode observed at low operating powers is predicted by the model.
3. The comparison holds for a time-averaged analysis of transient results, which mimics the way velocities are recorded by PIV.
4. The model is tractable since the Helmholtz equation is easier to solve than a system of ODEs representing bubble dynamics.

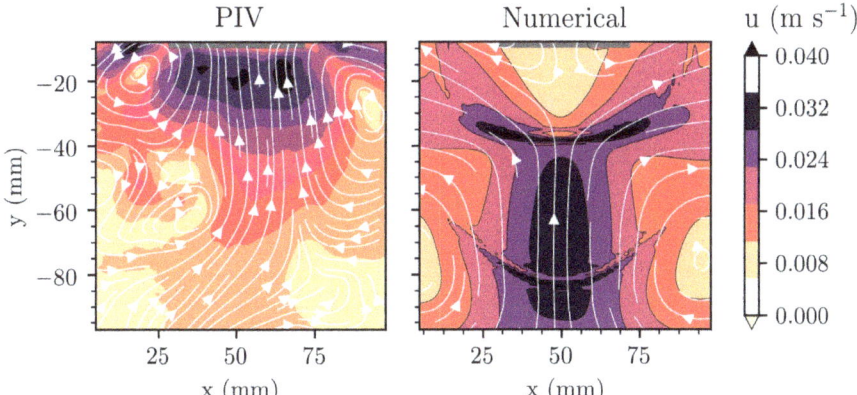

Figure 3. Comparison between measured velocities using particle image velocimetry (PIV) [63] in water and predictions of acoustic streaming using the numerical model described in [35]. The velocities are in m s^{-1}. The grey bar at the top of each contour represents the vibrating surface. The dataset used to reproduce these results is available elsewhere [69].

However, a remaining challenge is the suitable choice of the bubble number density as a parameter for the simulation. This study of acoustic streaming in water revealed that the predicted flow field is sensitive to the assumed bubble volume fraction, and therefore of bubble density by extension. The bubble density can also vary spatially, although a step function such as the one described by Equation (3) helps in limiting the presence of bubbles in regions below the Blake threshold. Semi-empirical values in the range 10^9–10^{12} m^{-3} have been reported to lead to acceptable results [32,35,54,70–72]. Aside of the difficulties in choosing the appropriate bubble density for a particular simulation, variation in bubble density due to differences in melt quality, presence of impurities and agglomerates or variation in degassing times presents further challenges.

This model was further improved using Trujillo's mathematically rigorous derivation of the complex wavenumber (9) and was then applied to DC casting using a continuum model [38]. Figure 4 shows the results of a numerical model of ultrasonic processing in DC casting. The predicted sump profile is altered by the fast streaming jet of hot liquid aluminium down the axis of the billet, thereby shortening the transition region at the centre of the billet. Observed grain orientations at the centre of cast billets confirm this prediction [38]. The model also predicts a slightly increased rate of porosity defects at the centre of billet cast with ultrasonic processing, as shown by the increased Niyama criterion at the centre of the domain (Figure 5). However, any experimental observation of increased porosity has not been reported so far.

Another research group independently used the same acoustic pressure formulation: Yamamoto and Komarov applied a similar model to aluminium [37] to reveal that attenuation of ultrasound and wavenumbers are larger in the molten aluminium than in water and that acoustic streaming flow is slower in the aluminium melt as compared with water. This is in agreement with our experimental study in [40], where shielding and acoustic damping were found to be more pronounced in liquid aluminium compared to water, obstructing the wave propagation into the bulk.

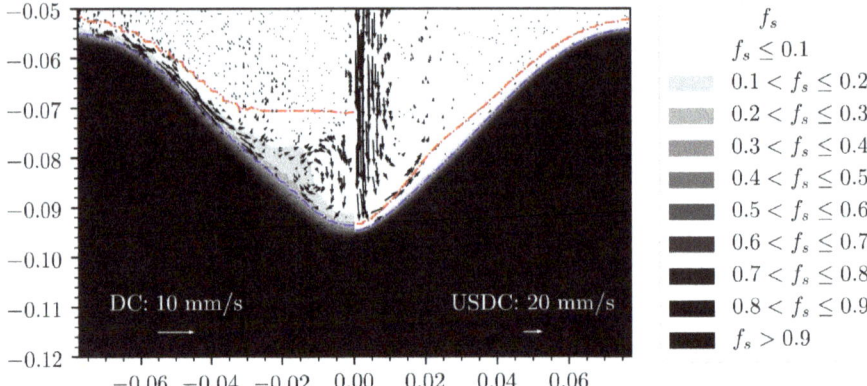

Figure 4. Comparison of sump profiles between conventional DC casting (left) and ultrasonic-assisted DC casting (right). f_s is the solid fraction using the casting conditions defined in [38]. Arrows are shown for the scale of the velocity field. The red dash-dotted line represents the liquidus temperature and the blue dash-dotted line denotes the coherency temperature (solid packing fraction). The dataset used to reproduce these results is available elsewhere [73].

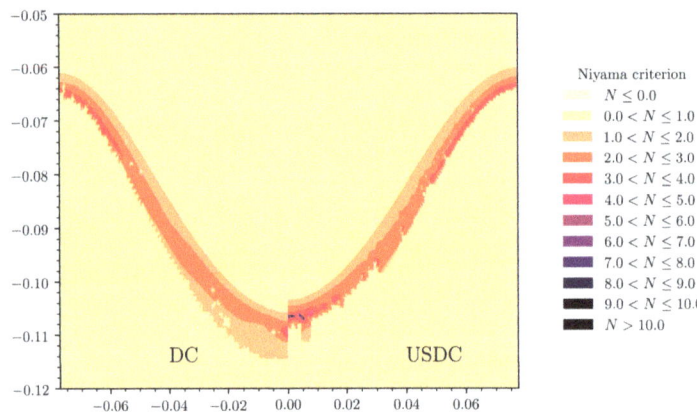

Figure 5. Comparison using the Niyama criterion between conventional DC casting (left) and ultrasonically assisted DC casting (right) using the casting conditions defined in [38]. The larger values upon sonication indicate an increased probability of porosity defects at the centre of the cast billet.

5. Current Challenges and Future Outlook

This summary has highlighted the requirement of considering the appropriate physics when modelling the complex phenomenon of ultrasonic melt treatment. However, more effort is required for an accurate prediction of actual treatment conditions. Aside from a recent contribution studying resonance in crucibles [58], the boundary conditions used in the models encountered in USP modelling so far are basic and do not take into account the vibration of the solid walls of a sonoreactor or reflection off real, rough crucible walls. Crevices in the walls could also act as seeds for nucleating bubbles and these are not taken into account by any model encountered so far. This is crucial in situations where resonance is required and both the changing bubbly media and imperfect container walls affect the resonant frequency of the system [58,74].

The stability of the heat balance solver upon mesh deformation and solidification front motion is still an issue for accurate modelling of ultrasonic processing in the presence of solidification. This limits the accuracy of casting simulations and need to be addressed for more accurate sump profile predictions. There is also uncertainty on the effect of the entrained cavitating bubbles and acoustic streaming jet on the packing fraction in the semi-solid region of a casting domain, and therefore on the delimitation between slurry and mushy zones. These need to be quantified accurately for more reliable predictions.

Since acoustic streaming modifies the grain morphology of the billet, the coherency temperature is expected to vary locally in the sump. Further study is required to determine the dependency of this parameter on the flow. However, an accurate a priori prediction is rendered difficult since the knowledge of the grain size and morphology is required before choosing the correct solid packing fraction.

Author Contributions: Conceptualization, B.L., I.T., K.P. and D.E.; Methodology, B.L. and I.T.; Software, B.L.; Validation, B.L. and I.T.; Formal Analysis, B.L. and I.T.; Investigation, B.L. and I.T.; Resources, B.L. and I.T.; Data Curation, B.L. and I.T.; Writing—original draft preparation, B.L.; Writing—review and editing, B.L., I.T., K.P. and D.E.; Visualization, B.L. and I.T.; Supervision, K.P. and D.E.; Project Administration, I.T., K.P. and D.E.; Funding Acquisition, I.T., K.P. and D.E.

Funding: This research as funded by the Engineering and Physical Sciences Research Council (EPSRC), UK, grant numbers EP/N007638/1, EP/R011001/1, EP/R011044/1 and EP/R011095/1.

Conflicts of Interest: The authors declare no conflict of interest.

References

1. Hill, M. Product and process design for structured products. *AIChE J.* **2004**, *50*, 1656–1661. [CrossRef]
2. Kowalski, A.J.; Cooke, M.; Hall, S. Expression for turbulent power draw of an in-line Silverson high shear mixer. *Chem. Eng. Sci.* **2011**, *66*, 241–249. [CrossRef]
3. Håkansson, A. Rotor-Stator Mixers: From Batch to Continuous Mode of Operation—A Review. *Processes* **2018**, *6*, 32. [CrossRef]
4. Eskin, D.G.; Tzanakis, I.; Wang, F.; Lebon, G.S.B.; Subroto, T.; Pericleous, K.; Mi, J. Fundamental studies of ultrasonic melt processing. *Ultrason. Sonochem.* **2019**, *52*, 455–467. [CrossRef] [PubMed]
5. Abramov, O.V. Action of high intensity ultrasound on solidifying metal. *Ultrasonics* **1987**, *25*, 73–82. [CrossRef]
6. Campbell, J. Effects of vibration during solidification. *Int. Met. Rev.* **1981**, *26*, 71–108. [CrossRef]
7. Eskin, G.I.; Eskin, D.G. *Ultrasonic Treatment of Light Alloy Melts*, 2nd ed.; Taylor & Francis, CRC Press: Boca Raton, FL, USA, 2015; ISBN 978-1-4665-7798-5.
8. Neppiras, E.A. Acoustic cavitation. *Phys. Rep.* **1980**, *61*, 159–251. [CrossRef]
9. Tzanakis, I.; Eskin, D.G.; Georgoulas, A.; Fytanidis, D.K. Incubation pit analysis and calculation of the hydrodynamic impact pressure from the implosion of an acoustic cavitation bubble. *Ultrason. Sonochem.* **2014**, *21*, 866–878. [CrossRef]
10. Campos-Pozuelo, C.; Granger, C.; Vanhille, C.; Moussatov, A.; Dubus, B. Experimental and theoretical investigation of the mean acoustic pressure in the cavitation field. *Ultrason. Sonochem.* **2005**, *12*, 79–84. [CrossRef]
11. Eskin, D.G. *Physical Metallurgy of Direct Chill Casting of Aluminum Alloys*; CRC Press: Boca Raton, FL, USA, 2008; ISBN 978-0-429-14523-0.
12. Wang, G.; Croaker, P.; Dargusch, M.; McGuckin, D.; StJohn, D. Simulation of convective flow and thermal conditions during ultrasonic treatment of an Al-2Cu alloy. *Comput. Mater. Sci.* **2017**, *134*, 116–125. [CrossRef]
13. Shu, D.; Sun, B.; Mi, J.; Grant, P.S. A High-Speed Imaging and Modeling Study of Dendrite Fragmentation Caused by Ultrasonic Cavitation. *Metall. Mater. Trans. A* **2012**, *43*, 3755–3766. [CrossRef]
14. Tzanakis, I.; Xu, W.W.; Eskin, D.G.; Lee, P.D.; Kotsovinos, N. In situ observation and analysis of ultrasonic capillary effect in molten aluminium. *Ultrason. Sonochem.* **2015**, *27*, 72–80. [CrossRef] [PubMed]
15. Tzanakis, I.; Xu, W.W.; Lebon, G.S.B.; Eskin, D.G.; Pericleous, K.; Lee, P.D. In Situ Synchrotron Radiography and Spectrum Analysis of Transient Cavitation Bubbles in Molten Aluminium Alloy. *Phys. Procedia* **2015**, *70*, 841–845. [CrossRef]
16. Xu, W.W.; Tzanakis, I.; Srirangam, P.; Mirihanage, W.U.; Eskin, D.G.; Bodey, A.J.; Lee, P.D. Synchrotron quantification of ultrasound cavitation and bubble dynamics in Al–10Cu melts. *Ultrason. Sonochem.* **2016**, *31*, 355–361. [CrossRef] [PubMed]

17. Wang, F.; Eskin, D.; Mi, J.; Wang, C.; Koe, B.; King, A.; Reinhard, C.; Connolley, T. A synchrotron X-radiography study of the fragmentation and refinement of primary intermetallic particles in an Al-35 Cu alloy induced by ultrasonic melt processing. *Acta Mater.* **2017**, *141*, 142–153. [CrossRef]
18. Wang, B.; Tan, D.; Lee, T.L.; Khong, J.C.; Wang, F.; Eskin, D.; Connolley, T.; Fezzaa, K.; Mi, J. Ultrafast synchrotron X-ray imaging studies of microstructure fragmentation in solidification under ultrasound. *Acta Mater.* **2018**, *144*, 505–515. [CrossRef]
19. Tudela, I.; Sáez, V.; Esclapez, M.D.; Díez-García, M.I.; Bonete, P.; González-García, J. Simulation of the spatial distribution of the acoustic pressure in sonochemical reactors with numerical methods: A review. *Ultrason. Sonochem.* **2014**, *21*, 909–919. [CrossRef]
20. Louisnard, O. A viable method to predict acoustic streaming in presence of cavitation. *Ultrason. Sonochem.* **2017**, *35*, 518–524. [CrossRef]
21. Caflisch, R.E.; Miksis, M.J.; Papanicolaou, G.C.; Ting, L. Effective equations for wave propagation in bubbly liquids. *J. Fluid Mech.* **1985**, *153*, 259. [CrossRef]
22. Wijngaarden, L.V. On the equations of motion for mixtures of liquid and gas bubbles. *J. Fluid Mech.* **1968**, *33*, 465. [CrossRef]
23. Foldy, L.L. The Multiple Scattering of Waves. I. General Theory of Isotropic Scattering by Randomly Distributed Scatterers. *Phys. Rev.* **1945**, *67*, 107–119. [CrossRef]
24. Atchley, A.A. The Blake threshold of a cavitation nucleus having a radius-dependent surface tension. *J. Acoust. Soc. Am.* **1989**, *85*, 152. [CrossRef]
25. Louisnard, O. A simple model of ultrasound propagation in a cavitating liquid. Part I: Theory, nonlinear attenuation and traveling wave generation. *Ultrason. Sonochem.* **2012**, *19*, 56–65. [CrossRef] [PubMed]
26. Keller, J.B.; Miksis, M. Bubble oscillations of large amplitude. *J. Acoust. Soc. Am.* **1980**, *68*, 628–633. [CrossRef]
27. Jamshidi, R.; Brenner, G. Dissipation of ultrasonic wave propagation in bubbly liquids considering the effect of compressibility to the first order of acoustical Mach number. *Ultrasonics* **2013**, *53*, 842–848. [CrossRef] [PubMed]
28. Gadi Man, Y.A.; Trujillo, F.J. A new pressure formulation for gas-compressibility dampening in bubble dynamics models. *Ultrason. Sonochem.* **2016**, *32*, 247–257. [CrossRef] [PubMed]
29. Toegel, R.; Gompf, B.; Pecha, R.; Lohse, D. Does Water Vapor Prevent Upscaling Sonoluminescence? *Phys. Rev. Lett.* **2000**, *85*, 3165–3168. [CrossRef]
30. Trujillo, F.J. A strict formulation of a nonlinear Helmholtz equation for the propagation of sound in bubbly liquids. Part I: Theory and validation at low acoustic pressure amplitudes. *Ultrason. Sonochem.* **2018**, *47*, 75–98. [CrossRef]
31. Commander, K.W.; Prosperetti, A. Linear pressure waves in bubbly liquids: Comparison between theory and experiments. *J. Acoust. Soc. Am.* **1989**, *85*, 732–746. [CrossRef]
32. Lebon, G.S.B.; Tzanakis, I.; Pericleous, K.; Eskin, D. Experimental and numerical investigation of acoustic pressures in different liquids. *Ultrason. Sonochem.* **2018**, *42*, 411–421. [CrossRef]
33. Eckart, C. Vortices and Streams Caused by Sound Waves. *Phys. Rev.* **1948**, *73*, 68–76. [CrossRef]
34. Zarembo, L.K. Acoustic Streaming. In *High-Intensity Ultrasonic Fields*; Rozenberg, L.D., Ed.; Springer US: Boston, MA, USA, 1971; pp. 135–199. ISBN 978-1-4757-5410-0.
35. Lebon, G.S.B.; Tzanakis, I.; Pericleous, K.; Eskin, D.; Grant, P.S. Ultrasonic liquid metal processing: The essential role of cavitation bubbles in controlling acoustic streaming. *Ultrason. Sonochem.* **2019**, *55*, 243–255. [CrossRef] [PubMed]
36. Fang, Y.; Yamamoto, T.; Komarov, S. Cavitation and acoustic streaming generated by different sonotrode tips. *Ultrason. Sonochem.* **2018**, *48*, 79–87. [CrossRef] [PubMed]
37. Yamamoto, T.; Komarov, S. Investigation on Acoustic Streaming During Ultrasonic Irradiation in Aluminum Melts. In *Light Metals 2019*; Chesonis, C., Ed.; Springer International Publishing: New York, NY, USA, 2019; pp. 1527–1531. ISBN 978-3-030-05863-0.
38. Lebon, G.S.B.; Salloum-Abou-Jaoude, G.; Eskin, D.; Tzanakis, I.; Pericleous, K.; Jarry, P. Numerical modelling of acoustic streaming during the ultrasonic melt treatment of direct-chill (DC) casting. *Ultrason. Sonochem.* **2019**, *54*, 171–182. [CrossRef] [PubMed]
39. Tzanakis, I.; Hodnett, M.; Lebon, G.S.B.; Dezhkunov, N.; Eskin, D.G. Calibration and performance assessment of an innovative high-temperature cavitometer. *Sens. Actuators A Phys.* **2016**, *240*, 57–69. [CrossRef]

40. Tzanakis, I.; Lebon, G.S.B.; Eskin, D.G.; Pericleous, K.A. Characterisation of the ultrasonic acoustic spectrum and pressure field in aluminium melt with an advanced cavitometer. *J. Mater. Process. Technol.* **2016**, *229*, 582–586. [CrossRef]
41. Nastac, L. Mathematical Modeling of the Solidification Structure Evolution in the Presence of Ultrasonic Stirring. *Metall. Mater. Trans. B* **2011**, *42*, 1297–1305. [CrossRef]
42. Nastac, L. Multiscale Modeling of Ingot Solidification Structure Controlled by Electromagnetic and Ultrasonic Stirring Technologies. In *CFD Modeling and Simulation in Materials Processing*; Nastac, L., Zhang, L., Thomas, B.G., Sabau, A., El-Kaddah, N., Powell, A.C., Combeau, H., Eds.; John Wiley & Sons, Inc.: Hoboken, NJ, USA, 2012; pp. 261–268. ISBN 978-1-118-36469-7.
43. Nastac, L. Numerical Modeling of Fluid Flow and Solidification Characteristics of Ultrasonically Processed A356 Alloys. *ISIJ Int.* **2014**, *54*, 1830–1835. [CrossRef]
44. Song, S.; Zhou, X.; Li, L.; Ma, W. Numerical simulation and experimental validation of SiC nanoparticle distribution in magnesium melts during ultrasonic cavitation based processing of magnesium matrix nanocomposites. *Ultrason. Sonochem.* **2015**, *24*, 43–54. [CrossRef]
45. Lebon, G.S.B.; Pericleous, K.; Tzanakis, I.; Eskin, D. A model of cavitation for the treatment of a moving liquid metal volume. *Int. J. Cast Met. Res.* **2016**, *29*, 324–330. [CrossRef]
46. Singhal, A.K.; Athavale, M.M.; Li, H.; Jiang, Y. Mathematical Basis and Validation of the Full Cavitation Model. *J. Fluid. Eng.* **2002**, *124*, 617. [CrossRef]
47. Plesset, M.S. The dynamics of cavitation bubbles. *J. Appl. Mech.* **1949**, *16*, 277–282. [CrossRef]
48. Neppiras, E.A.; Noltingk, B.E. Cavitation Produced by Ultrasonics: Theoretical Conditions for the Onset of Cavitation. *Proc. Phys. Soc. Sect. B* **1951**, *64*, 1032–1038. [CrossRef]
49. Jasper, J.J. The Surface Tension of Pure Liquid Compounds. *J. Phys. Chem. Ref. Data* **1972**, *1*, 841–1010. [CrossRef]
50. Chen, Y.-J.; Hsu, W.-N.; Shih, J.-R. The Effect of Ultrasonic Treatment on Microstructural and Mechanical Properties of Cast Magnesium Alloys. *Mater. Trans.* **2009**, *50*, 401–408. [CrossRef]
51. Puga, H.; Barbosa, J.; Carneiro, V.H. The Role of Acoustic Pressure during Solidification of AlSi7Mg Alloy in Sand Mold Casting. *Metals* **2019**, *9*, 490. [CrossRef]
52. Wang, S.; Kang, J.; Guo, Z.; Lee, T.L.; Zhang, X.; Wang, Q.; Deng, C.; Mi, J. In situ high speed imaging study and modelling of the fatigue fragmentation of dendritic structures in ultrasonic fields. *Acta Mater.* **2019**, *165*, 388–397. [CrossRef]
53. Lebon, G.S.B.; Tzanakis, I.; Pericleous, K.A.; Eskin, D.G. Comparison between low-order and high-order acoustic pressure solvers for bubbly media computations. *J. Phys. Conf. Ser.* **2015**, *656*, 012134. [CrossRef]
54. Lebon, G.S.B.; Tzanakis, I.; Djambazov, G.; Pericleous, K.; Eskin, D.G. Numerical modelling of ultrasonic waves in a bubbly Newtonian liquid using a high-order acoustic cavitation model. *Ultrason. Sonochem.* **2017**, *37*, 660–668. [CrossRef]
55. Manoylov, A.; Lebon, B.; Djambazov, G.; Pericleous, K. Coupling of Acoustic Cavitation with DEM-Based Particle Solvers for Modeling De-agglomeration of Particle Clusters in Liquid Metals. *Met. Mater. Trans. A* **2017**, *48*, 5616–5627. [CrossRef]
56. Lebon, G.S.B.; Kao, A.; Pericleous, K. The uncertain effect of cavitating bubbles on dendrites. In Proceedings of the 6th Decennial International Conference on Solidification Processing, Old Windsor, UK, 25–28 July 2017; pp. 554–557.
57. Wang, Y.; Lebon, B.; Tzanakis, I.; Zhao, Y.; Wang, K.; Stella, J.; Poirier, T.; Darut, G.; Liao, H.; Planche, M.-P. Experimental and numerical investigation of cavitation-induced erosion in thermal sprayed single splats. *Ultrason. Sonochem.* **2019**, *52*, 336–343. [CrossRef] [PubMed]
58. Tonry, C.E.H.; Djambazov, G.; Dybalska, A.; Bojarevics, V.; Griffiths, W.D.; Pericleous, K.A. Resonance from Contactless Ultrasound in Alloy Melts. In *Light Metals 2019*; Chesonis, C., Ed.; Springer International Publishing: New York, NY, USA, 2019; pp. 1551–1559. ISBN 978-3-030-05863-0.
59. Djambazov, G.S.; Lai, C.-H.; Pericleous, K.A. Staggered-Mesh Computation for Aerodynamic Sound. *AIAA J.* **2000**, *38*, 16–21. [CrossRef]
60. Huang, H.; Shu, D.; Fu, Y.; Zhu, G.; Wang, D.; Dong, A.; Sun, B. Prediction of Cavitation Depth in an Al-Cu Alloy Melt with Bubble Characteristics Based on Synchrotron X-ray Radiography. *Met. Mater. Trans. A* **2018**, *49*, 2193–2201. [CrossRef]

61. Jamshidi, R.; Pohl, B.; Peuker, U.A.; Brenner, G. Numerical investigation of sonochemical reactors considering the effect of inhomogeneous bubble clouds on ultrasonic wave propagation. *Chem. Eng. J.* **2012**, *189–190*, 364–375. [CrossRef]
62. Chopra, K.; Prasada Rao, A.K. Ultrasonicated Direct Chill Clad Casting of Magnesium Alloy: A Computational Approach. *Trans. Indian Inst. Met.* **2019**, 1–7. [CrossRef]
63. Tzanakis, I.; Lebon, G.S.B.; Eskin, D.; Hyde, M.; Grant, P.S. Investigation of acoustic streaming and cavitation intensity in water as an analogue for liquid metal. In Proceedings of the 10th International Symposium on Cavitation (CAV2018), Baltimore, MD, USA, 14–16 May 2018.
64. Wang, G.; Croaker, P.; Dargusch, M.; McGuckin, D.; StJohn, D. Evolution of the As-Cast Grain Microstructure of an Ultrasonically Treated Al-2Cu Alloy. *Adv. Eng. Mater.* **2018**, *20*, 1800521. [CrossRef]
65. Wang, G.; Wang, Q.; Balasubramani, N.; Qian, M.; Eskin, D.G.; Dargusch, M.S.; StJohn, D.H. The Role of Ultrasonically Induced Acoustic Streaming in Developing Fine Equiaxed Grains During the Solidification of an Al-2 Pct Cu Alloy. *Met. Mater. Trans. A* **2019**. [CrossRef]
66. Lee, Y.K.; Youn, J.I.; Kim, Y.J. Modeling of the Effect of Ultrasonic Frequency and Amplitude on Acoustic Streaming. In *Light Metals 2019*; Chesonis, C., Ed.; Springer International Publishing: New York, NY, USA, 2019; pp. 1573–1578. ISBN 978-3-030-05863-0.
67. Lee, Y.K.; Youn, J.I.; Hwang, J.H.; Kim, J.H.; Kim, Y.J.; Lee, T.Y. Modeling of the effect of ultrasonic amplitude and frequency on acoustic streaming. *Jpn. J. Appl. Phys.* **2019**, *58*, SGGD07. [CrossRef]
68. Riedel, E.; Horn, I.; Stein, N.; Stein, H.; Bähr, R.; Scharf, S. Ultrasonic treatment: A clean technology that supports sustainability in casting processes. *Procedia CIRP* **2019**, *80*, 101–107. [CrossRef]
69. Lebon, B.; Tzanakis, I.; Pericleous, K.; Eskin, D.; Grant, P.S. Ultrasonic Liquid Metal Processing: The Essential Role of Cavitation Bubbles in Controlling Acoustic Streaming Dataset. Available online: https://brunel.figshare.com/articles/Ultrasonic_liquid_metal_processing_The_essential_role_of_cavitation_bubbles_in_controlling_acoustic_streaming_dataset/7435457 (accessed on 1 October 2019).
70. Vanhille, C.; Campos-Pozuelo, C. Nonlinear Ultrasonic Propagation in Bubbly Liquids: A Numerical Model. *Ultrasound Med. Biol.* **2008**, *34*, 792–808. [CrossRef]
71. Vanhille, C.; Campos-Pozuelo, C. Nonlinear ultrasonic waves in bubbly liquids with nonhomogeneous bubble distribution: Numerical experiments. *Ultrason. Sonochem.* **2009**, *16*, 669–685. [CrossRef] [PubMed]
72. Vanhille, C.; Campos-Pozuelo, C. Nonlinear ultrasonic standing waves: Two-dimensional simulations in bubbly liquids. *Ultrason. Sonochem.* **2011**, *18*, 679–682. [CrossRef] [PubMed]
73. Lebon, B.; Salloum-Abou-Jaoude, G.; Eskin, D.; Tzanakis, I.; Pericleous, K.; Jarry, P. Numerical Modelling of Acoustic Streaming during the Ultrasonic Melt Treatment of Direct-Chill (DC) Casting Dataset. Available online: https://brunel.figshare.com/articles/Numerical_modelling_of_acoustic_streaming_during_the_ultrasonic_melt_treatment_of_direct-chill_DC_casting_dataset/7610924/1 (accessed on 1 October 2019).
74. Pericleous, K.A.; Bojarevics, V.; Djambazov, G.; Dybalska, A.; Griffiths, W.; Tonry, C. The Contactless Electromagnetic Sonotrode. In *Shape Casting*; Tiryakioğlu, M., Griffiths, W., Jolly, M., Eds.; Springer International Publishing: New York, NY, USA, 2019; pp. 239–252. ISBN 978-3-030-06033-6.

© 2019 by the authors. Licensee MDPI, Basel, Switzerland. This article is an open access article distributed under the terms and conditions of the Creative Commons Attribution (CC BY) license (http://creativecommons.org/licenses/by/4.0/).

Article

Characterization of Ultrasonic Bubble Clouds in A Liquid Metal by Synchrotron X-ray High Speed Imaging and Statistical Analysis

Chuangnan Wang [1], Thomas Connolley [2], Iakovos Tzanakis [3], Dmitry Eskin [4] and Jiawei Mi [1,*]

1. Department of Engineering, University of Hull, Cottingham Road, Hull HU6 7RX, UK; Chuangnan.wang@ionix.at
2. Diamond Light Source, Didcot OX11 0DE, Oxfordshire, UK; thomas.connolley@diamond.ac.uk
3. Oxford Brooks University, Oxford OX33 1HX, Oxfordshire, UK; itzanakis@brookes.ac.uk
4. Brunel Centre for Advanced Solidification Technology, Brunel University London, Uxbridge UB8 3PH, UK; Dmitry.Eskin@brunel.ac.uk
* Correspondence: J.Mi@hull.ac.uk

Received: 18 November 2019; Accepted: 16 December 2019; Published: 20 December 2019

Abstract: Quantitative understanding of the interactions of ultrasonic waves with liquid and solidifying metals is essential for developing optimal processing strategies for ultrasound processing of metal alloys in the solidification processes. In this research, we used the synchrotron X-ray high-speed imaging facility at Beamline I12 of the Diamond Light Source, UK to study the dynamics of ultrasonic bubbles in a liquid Sn-30wt%Cu alloy. A new method based on the X-ray attenuation for a white X-ray beam was developed to extract quantitative information about the bubble clouds in the chaotic and quasi-static cavitation regions. Statistical analyses were made on the bubble size distribution, and velocity distribution. Such rich statistical data provide more quantitative information about the characteristics of ultrasonic bubble clouds and cavitation in opaque, high-temperature liquid metals.

Keywords: ultrasonic bubble clouds; synchrotron X-ray imaging; metal solidification; ultrasound melt processing

1. Introduction

Ultrasonic cavitation created by high-power ultrasound in liquids is a highly dynamic and nonlinear process. It has very wide applications in industry, for example, ultrasound cleaning [1], sonochemistry [2], metallurgy [3,4]. Historically, cavitation was first studied by Lord Rayleigh in the late 19th century, when he considered the collapse of a spherical void within a liquid [4,5]. During and after the 2nd World War, cavitation in water was studied extensively in the field of hydrodynamics, because it was a very common phenomenon that significantly affected the efficiency of pumps and propellers used for ships [6,7]. For ultrasonic cavitation, many invasive and non-invasive experimental methods have been developed for measuring and characterizing ultrasonic cavitation zone and its intensity. For example, acoustic hydrophones, cavitation meters and foil testing are typical invasive techniques. The probe/sensor or testing foil is inserted into the ultrasonic cavitation region to measure the characteristics of the cavitation zone, its size, position and intensity [8–10]. Non-invasive techniques include ultrasonic phase array transducers with ultrasonic imaging, visible light imaging [8,11,12] and Optical Diffraction Tomography (ODT) [13,14]. In non-invasive measurement techniques, no probes are used to interfere with the acoustic/cavitation field.

Photography is a common method for measuring ultrasound cavitation in light transparent liquids. For example, luminol mapping can measure cavitation intensity distribution directly [15,16] based on the brilliance of the emitted light in the cavitation process. ODT uses the diffraction of light by ultrasound and a tomographic technique to form images for the pressure in a plane perpendicular to the ultrasound propagation direction. However, to study the highly dynamic behavior of cavitation, higher temporal and spatial resolution are needed. Hence, high-speed cameras with a higher magnification objective lens is often used [8,17]. For example, Geisler studied bubble oscillation using an acquisition rate of 2 million frames per second in a view field of 160 µm × 160 µm. While Ohl investigated bubble collapse in water using an exposure time of 48 ns, but a single-shot image in a view field of 1.5 mm × 1.8 mm [8,17]. These high-speed imaging studies resulted in significant progress on the understanding of bubble oscillation, bubble shock wave emission, bubble luminescence and liquid flow in the vicinity of bubbles. However, in the cases of opaque, high-viscosity and high-temperature materials such as liquid metals, the methods described above are not suitable.

An in-depth understanding of the interactions of ultrasonic waves with liquid and solidifying metals is essential for developing optimal processing strategies for ultrasound processing of metal alloys [4,18,19]. Recently, researchers in Mi's group have extensively studied the dynamics of ultrasonic bubbles in liquid metals and their interactions with the growing phases in the liquid, and with the solid-liquid interface [20–24]. These studies have provided real-time and convincing evidence to clarify that (1) the shock wave created at bubble implosion, and (2) the cyclic fatigue effects due to bubble oscillation are the most important mechanisms for the fragmentation of growing dendritic structures and intermetallic phases during the metal solidification processes. High-power ultrasound normally produces a large amount of bubbles in liquid metals (often called a bubble cloud). So far, a quantitative characterisation of those bubble clouds in liquid metals have not been reported. In this aspect, real-time and quantitative studies of ultrasonic bubble clouds in liquid metal can provide more in-depth understanding of the characteristics of ultrasonic bubble clouds.

2. Experiments

2.1. Alloy and Sample Preparation

A Sn-30wt%Cu alloy was chosen as the experimental alloy. It has a wide solidification range, allowing ultrasound to be applied over a wide range of temperatures (from 250 to 750 °C). Furthermore, the Sn–Cu binary system [25] is the key alloy system for lead-free soldering materials, and intermetallic compounds such as Cu_6Sn_5 are promising candidates for enhancing the storage capacity of Lithium ion-based batteries. Samples were contained inside specially made quartz capsules with a flattened "hourglass" shape, as shown in Figure 1. The thin central window section of the capsule was 300 µm thick to allow good X-ray penetration. The sonotrode was positioned in the upper end of the hourglass, with the tip close to the top of the thin section. A heat sink made of a stainless steel rod was placed at the bottom reservoir of the hourglass capsule to create a thermal gradient. The alloy in the capsule was melted inside a small cartridge heater furnace [26]. Sample temperature was monitored and recorded by three K-type thermocouples placed at the top, middle and bottom of the thin window section. The details of the experiment are very similar to those described in [22,24].

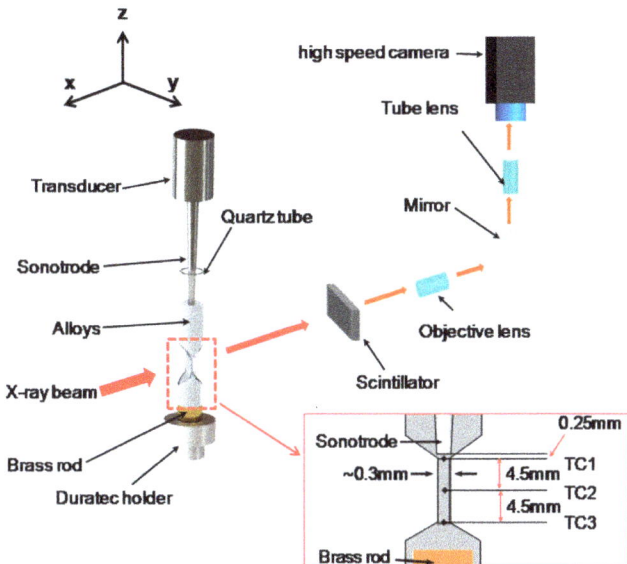

Figure 1. A 3D CAD rendition, showing the experimental set-up; and a 2D sketch of the structure of sample holder, sonotrode and thermal couple (TC 1, 2 and 3) position.

2.2. High-Speed Synchrotron X-ray Imaging

The experiment was carried out at Beamline I12, Diamond Light Source, Oxford, UK [27], with the set up illustrated in Figure 1.

I12 has a wiggler source that delivers a peak flux of ~1.7 × 10^{11} photons/s/mm²/0.1%BW in the first experimental hutch. Filtered white beam was used to give maximum X-ray flux on the sample for high-speed image acquisition. Filtration of the beam (4 mm of copper) was necessary to reduce heat load on the sample and reduce the risk of damage to the X-ray detector. The X-ray camera system consisted of a 200 μm thick LYSO scintillator lens coupled to a Miro 310M high-speed CMOS camera operating at a 1280 × 800 pixel image size with a resolution of 4 μm × 4 μm per pixel. Under these conditions with the given samples, frame rates of 2000 frames per second (fps) were obtained, using an exposure of 499 μs, for a total recording time of 4 s. The recording time was limited by the camera's fast onboard memory buffer (12GB). However, this was adequate for observing the ultrasonic phenomena in the experiments. Flat field images without a sample were recorded to enable correction for non-uniformity in the incident beam intensity profile and imperfections in the imaging system.

Once the desired target temperature (675 °C) was reached and stabilized in the liquid metal, a TTL trigger unit was used to initiate recording of images using a high-speed X-ray camera. After a ~10 ms delay, ultrasound was applied to the liquid metal. The ultrasound was generated using an UP100H ultrasonic processor with an MS2 sonotrode (Hielscher ultrasound technology Ltd., Teltow, Germany). Ultrasound powers of 20 W, 60 W and 100 W were used in the experiment to create bubble clouds with different characteristics. In this way, the whole process of ultrasound bubble nucleation, growth and propagation can be captured.

Two X-ray videos were taken for each ultrasonic power setting. The first video was taken of the area just below the sonotrode tip, with the aim of studying cavitation phenomena close to the tip. In this paper, the location is referred to as chaotic cavitation region (CCR). A second video was taken 1.5 mm below the first set of images. This second video was targeted at phenomena which happen further away from the sonotrode tip, referred to as the quasi static cavitation region (SCR) in this paper.

3. Image Processing and Data Analysis

Flat-field images were taken without any sample between the X-ray source and detector, as shown in Figure 2a. The images were then filtered using the Remove Outlier function in ImageJ [28] to remove random bright outlier pixels caused by direct X-ray strikes on the camera sensor. Such strikes are the consequence of the high-intensity, high-energy beam and cannot totally be eliminated by shielding of the camera system. The acquired videos were flat-field corrected by dividing the flat-field image to remove the effect of non-uniform X-ray beam intensity and compensate for systematic variations in the detector system, such as vignetting, or dust or scratches on the scintillator. Dark count and bad pixel correction were also performed by the camera hardware. Typical images obtained after flat-field correction and filtering are presented in Figure 2b,c.

Table 1. Grey level and thickness of the three locations, g_0, g_1 and g_2 shown in Figure 2.

Position	Thickness of Liquid Metal (mm)	Grey Level
g_0	0	145
g_1	2	89
g_2	0.3	97

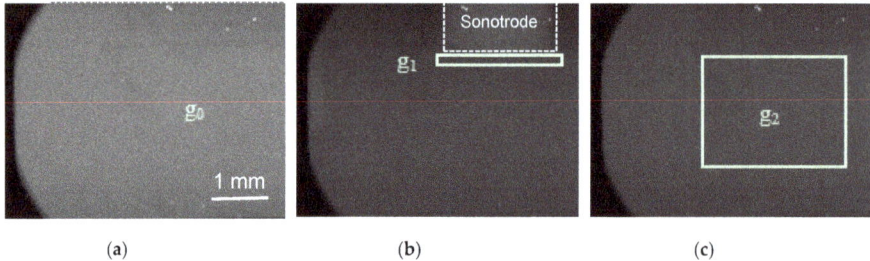

Figure 2. X-ray images of liquid Sn-30wt%Cu alloy, without the application of ultrasound, (**a**) a typical flat-field image, and the black regions at the corners of the images are from the lens mount of the camera. (**b**) in the chaotic cavitation region with the sonotrode (marked with the white dotted line) in the view field; (**c**) in the quasi static cavitation region, i.e., 1.5 mm below the sonotrode. g_0, g_1 and g_2 mark the locations (including the grey level) where the thicknesses of the liquid metal are known as listed in Table 1.

In the experiments, high-speed images sequences in the CCR and SCR regions were captured under different ultrasound powers. In the CCR, the ultrasonic bubbles are highly dynamic and interconnected. In the SCR, individual ultrasonic bubbles were clearly seen. As bubbles attenuate much less X-rays than the liquid Sn-30wt%Cu alloy, they had a higher grey level in images than the surrounding liquid alloy and can be segmented, and individually counted and measured. To quantify the different cavitation characteristics, different methods were used. Cavitation in the CCR was analysed based on grey-level differences, while cavitation in the SCR was analysed by counting and measuring individual bubbles.

3.1. Bubble Volume Fraction in the Chaotic Cavitation Region

In the CCR region, ultrasonic bubbles were highly dynamic and interconnected to form bubble clouds. It was impossible, at 2000 fps, to capture and distinguish the individual bubbles. To characterise the collective behaviour of the bubble clouds, the acquired images were analysed based on the grey-level distribution caused by the presence of the bubble clouds, because they have lower X-ray attenuation, and result in a brighter area in the images. If monochromatic (single-wavelength) X-rays were used, the transmitted intensity through the sample can be calculated analytically from a known X-ray mass attenuation coefficient μ using the Beer–Lambert Law [29]. For an incident intensity I_0, the transmitted intensity I through a material of density ρ and thickness d is given by:

$$I = I_0 e^{-\mu \rho d} \quad (1)$$

μ is a material property and it is a function of photon energy. However, in this study, monochromatic X-rays do not have sufficient flux to achieve the short exposure times associated with the high frame rate that is required in this experiment. Instead, broad spectrum multi-wavelength X-rays, known as white-beam X-rays, were used. For white-beam X-rays, the Beer–Lambert exponential attenuation law cannot be used directly, because both the mass attenuation coefficient of the sample and the X-ray detector response are photon-energy-dependent. Therefore, an approximation method based on grey-level measurement is adopted to calculate the bubble cloud volume fraction in this paper. The diameter of the sonotrode tip was 2 mm, which means that the thickness of the material just under the sonotrode tip, indicated by the rectangle in Figure 2b, was approximately 2 mm. As showed in the insert of Figure 1, the narrow tube thickness was 0.3 mm (the corresponding X-ray image through this area is illustrated in Figure 2c). The thicknesses at these two positions were used as the reference to calculate the overall attenuation coefficients. We assumed that the sample thickness and grey level are exponentially related. At both positions, the empirical attenuation coefficients were approximated by the equation below:

$$g = g_0 e^{-\alpha_g x + \beta_g} \quad (2)$$

Here, g_0 is the grey level of the corresponding position in the flat-field image, x is the transmitting distance and α_g and β_g are the empirical attenuation coefficients, representing the attenuation parameters in this case. Using the grey levels found at the locations of g_0, g_1 and g_2 (see Table 1), α_g and β_g are calculated as 124.2 and −0.33, respectively. Equation (2) was then used to calculate the thickness of the liquid Sn-30wt%Cu (i.e., x in Equation (2)). The line density along the X-ray propagation direction where ultrasonic bubbles were found can be calculated by using:

$$LD = (D_0 - D)/D_0 \quad (3)$$

where D_0 is the thickness of the liquid metal without ultrasonic bubbles, and D is the thickness of the sample containing ultrasonic bubbles. Both are calculated by Equation (2). For a unit volume, (a unit area perpendicular to the X-ray transmission direction times a unit length along the X-ray direction), the volume fractions of the bubbles in this unit volume at the CCR can be calculated using Equation (3) and are actually represented by LD.

3.2. Bubble Volume Fraction in the Quasi Static Cavitation Region

For the SCR region (1.5 mm below the region that contains the sonotrode), individual ultrasonic bubbles or bubble clouds were observed and recorded. To count these bubbles, the images were firstly normalized by the X-ray images that were taken without ultrasound treatment (UST) to remove any non-uniform grey level caused by the non-uniform thickness of the glass cell windows. Secondly, the contrast between bubbles and background was enhanced by using the inverse of a natural exponential function as demonstrated in Figure 3c. Thirdly, the image was converted into binary images based on the predefined grey-level threshold. Finally, the bubble dimensions and positions were determined using the MATLAB Image Processing Toolbox, which is illustrated in Figure 3d. In this way, a statistical analysis of ultrasonic bubbles was made and described in Section 4.

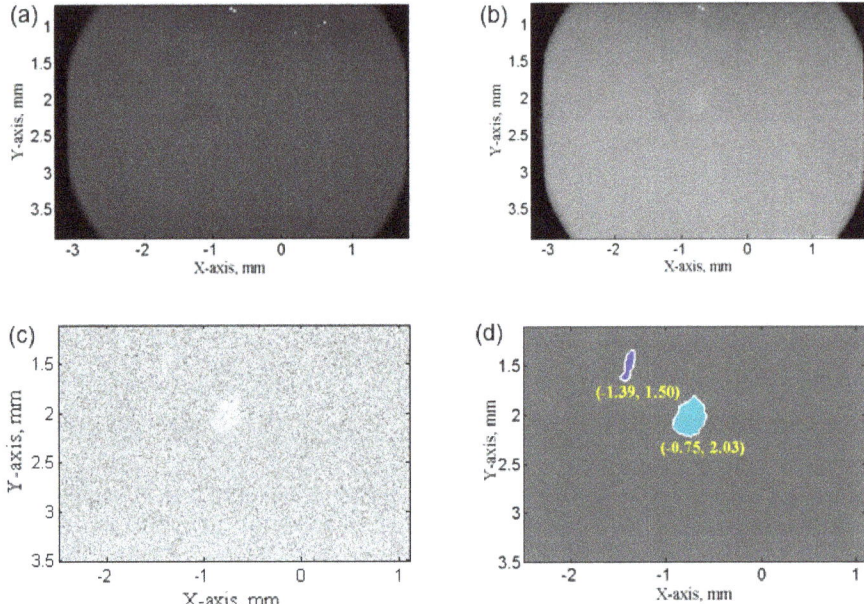

Figure 3. A typical X-ray raw image taken in the SCR region. (**a**) before image processing; (**b**) after flat-field correction; (**c**) cropped X-ray image after normalization by image taken without UST, and contrast enhancement; (**d**) The final binary image, which was used to extract the position and size of ultrasonic bubbles.

4. Experimental Results

4.1. Bubble Cloud in the Chaotic Cavitation Region

The ultrasound bubble clouds observed in the CCR, at different ultrasound powers (20 W, 60 W and 100 W, respectively) are illustrated in Figure 4. In each case, 10 sequential X-ray images were averaged to reduce background noise, so that each result in the figure contained the information averaged over 5 ms. Figure 4d–f, shows the bubble volume fraction calculated using Equation (3).

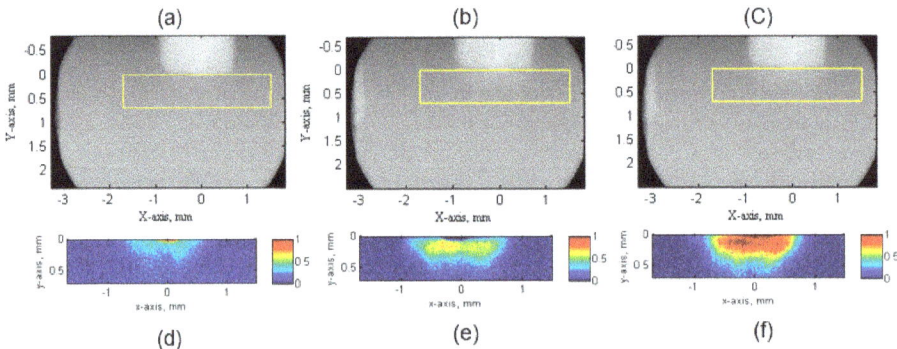

Figure 4. High-speed synchrotron X-ray images of ultrasonic bubble clouds near the sonotrode tip at different ultrasound powers of (**a**) 20 W, (**b**) 60 W, (**c**) 100 W respectively. (**d**–**f**) show the corresponding bubble volume fraction calculated based on Equation (3).

Figure 4d–f clearly show that the bubble volume fraction increases with increasing ultrasound power. Figure 5 further plots the line distribution characteristics in the horizontal and vertical directions. Figure 5b clearly shows that the maximum cavitation occurred near the sonotrode tip. Measurements of cavitation intensity and cavitation cloud dimensions are summarized in Table 2. Cavitation cloud width and length in the x and y direction are characterised by the full width at half maximum (FWHM). The measurement indicated that there was no significant variation in the physical dimensions of the cavitation cloud. However, cavitation intensity, which was evaluated in terms of the average bubble volume fraction, increased significantly with the increase of ultrasound power.

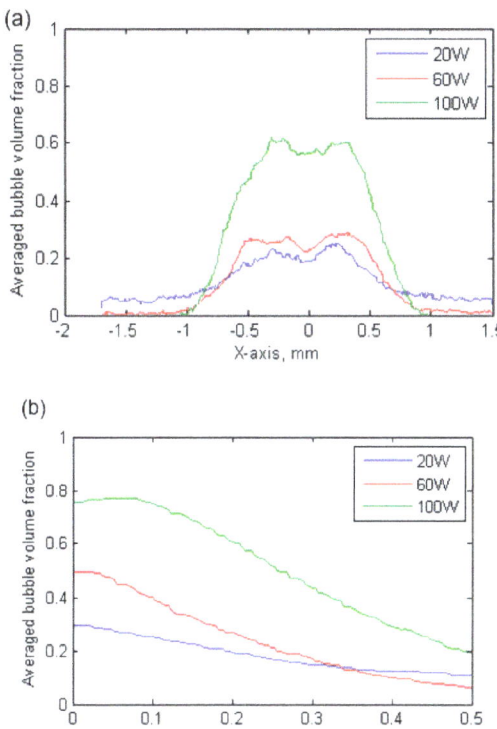

Figure 5. The averaged bubble volume fraction in (**a**) horizontal direction from position −1.5 mm to 1.5 mm (X-axis), to show bubble volume fraction distribution in horizontal direction; in (**b**) vertical direction from sonotrode tip to 0.5 mm (Y-axis), to show bubble volume fraction distribution in vertical direction.

Table 2. Ultrasonic bubble cloud characteristics due to different ultrasonic powers.

Ultrasound Power (W)	Ultrasonic Bubble Cloud * FWHM in X-axis (mm)	Ultrasonic Bubble Cloud FWHM in Y-axis (mm)	Maximum Averaged Bubble Volume Fraction in X-axis	Maximum Averaged Bubble Volume Fraction in Y-axis
20	1.15	0.30	0.22	0.31
60	1.26	0.30	0.38	0.52
100	1.26	0.34	0.71	0.78

* FWHM is the full width at half maximum.

4.2. Bubble Cloud in the Quasi-Static Cavitation Region

In the SCR, individual bubbles were observed and hence tracking and counting of the imaged bubbles are possible. MATLAB was employed to track and measure bubble sizes and velocities. The region of interest for the measurement is shown by the yellow box in Figure 6. The region of interest was further divided into three sub regions: Section 1 (1.2 mm to 2 mm, distance to sonotrode), Section 2 (2 mm to 2.8 mm, distance to sonotrode), Section 3 (2.8 mm to 3.6 mm, distance to sonotrode). A statistical study of bubbles was performed on a sequence of 2000 X-ray images (1 s duration of ultrasound processing) acquired in a quasi steady-state condition (1 s after the ultrasound was turned on, and a steady-state condition was established in the liquid alloy).

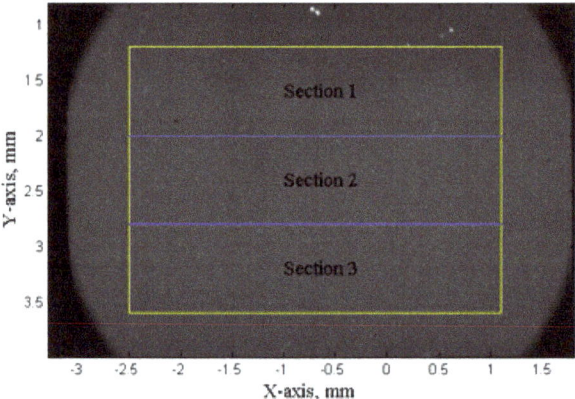

Figure 6. A typical X-ray image of the SCR and three sub-regions where data analysis was made.

4.2.1. Bubble Size Distribution

Figure 7 shows the bubble size distribution for different ultrasound powers. The Kernel probability density function [30] was used to calculate the probability densities of these bubble size distributions. The maximum probability density occurred at a bubble size of 0.0078 mm^2, 0.0098 mm^2 and 0.0096 mm^2 (in Table 3) for 20 W, 60 W and 100 W ultrasonic powers, respectively. These results indicate that bubbles with a diameter of approximately 0.01 mm^2 were dominant in the bubble population. The observed bubble size was mainly dependent on two factors: (1) the initial bubble radius (i.e., the nuclei size), which is related to the liquid properties (viscosity, density, etc.) and (2) ultrasound pressure magnitude and frequency. It should note that the camera resolution is 4 μm per pixel, which means that a bubble diameter smaller than 10 μm cannot be resolved.

The statistical results show that 20 W ultrasound power generated 16 bubbles with an area larger than 0.1 mm^2, which is 1.6% of its total bubble population. A 60 W ultrasound power generated 234 bubbles with an area larger than 0.1 mm^2, which is 9.1% of its total bubble population; a 100 W ultrasound power generated 1151 bubbles with an area larger than 0.1 mm^2, which is 17.5% of its total bubble population. Clearly, a higher ultrasound power resulted in a larger probability of bigger-sized bubbles.

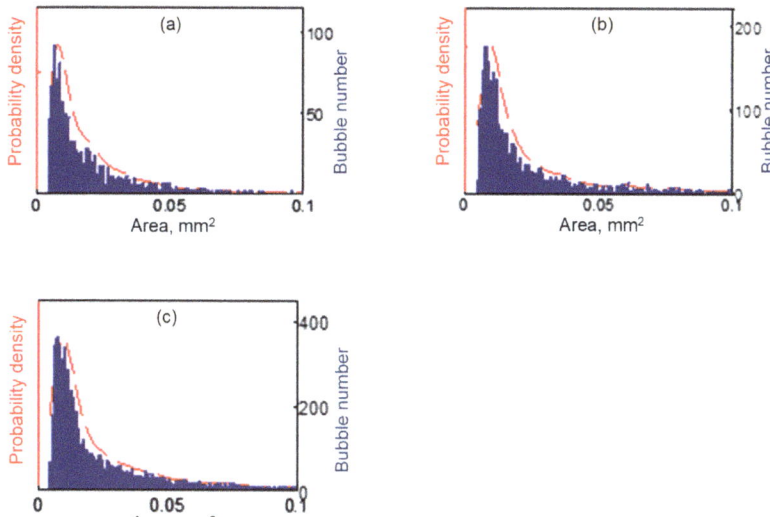

Figure 7. Ultrasonic bubble sizes distribution in the SCR with ultrasound power of (**a**) 20 W, (**b**) 60 W and (**c**) 100 W respectively.

Table 3. Ultrasonic bubble size and velocity distribution.

Ultrasound Power (W)	Bubble Size at Maximum Probability Density (mm^2)	Bubble Velocity at Maximum Probability Density (m/s)
20	0.0078	0.15 m/s
60	0.0098	0.28 m/s
100	0.0096	0.17 m/s

4.2.2. Bubble Velocity Distribution

Individual bubble velocities were calculated by comparing the position of a bubble centroid in consecutive images. The distances between the centroids of the same bubble in consecutive frames were taken as the distance travelled in the interframe time of the image sequence. The velocities of bubbles were calculated by dividing the measured distance during the interframe time. By measuring the bubble velocities in a one second image sequence, velocity distributions were obtained and are plotted as histograms in Figure 8 with the probability densities plotted as dashed lines. The velocities corresponding to the maximum probability density were 0.15 m/s, 0.28 m/s and 0.17 m/s, for 20 W, 60 W and 100 W ultrasonic powers, respectively. The most probable velocity at 100 W (0.17 m/s) was lower than that at 60 W. We think this may be due to the imaging method we used. At 100 W power, some bubble velocities were too fast for the same bubble to be captured in two consecutive images. This means that some fast-moving bubbles are missing from the bubble velocity datasets, and statistical analysis was only performed on bubbles that were moving slowly enough to be captured in two consecutive frames.

Imaging at higher frame rates would be able to measure the velocity of fast-moving bubbles in the 100 W case. However, with the given camera sensitivity, sample and X-ray source characteristics, the signal/noise ratio in even higher-speed images would be lower, making bubble identification more difficult.

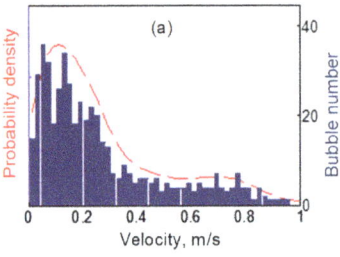
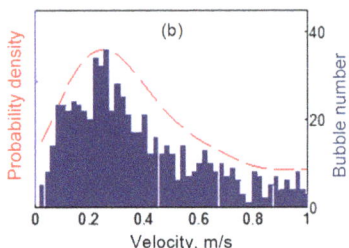
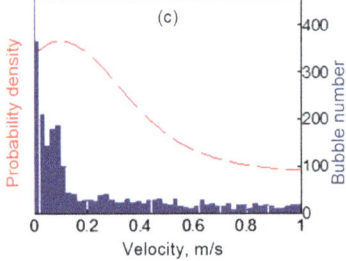

Figure 8. Bubble velocity distributions in the SCR at ultrasound powers of (**a**) 20 W, (**b**) 60 W, (**c**) 100 W, respectively.

4.3. Limitation of the Imaging Method

In this research, we acquired many terabytes of image sequences. To extract useful information from these large-scale image datasets, a simple but robust method had to be developed. In our case, ultrasound bubbles away from the sonotrode, which we refer it as SCR, were identified based on the grey level. It is interesting to see that, in the field of view of the images taken in this research, the bubble size distribution away from the sonotrode does not change significantly as the ultrasound power increases, although normally, a higher acoustic pressure would lead to larger bubbles. For characterising the bubbles, the imaging method was limited by the acquisition frame rate (2000 fps), especially for the case of 100 W, where bubbles can travel out of the view field in two consecutive images, making the tracking of the exact location of the bubble more difficult.

In our experiment, the ultrasound cavitation bubble cloud is found to be in a relative stable shape, and it is due to the higher ultrasound attenuation of the liquid metal.

5. Conclusions

Synchrotron X-ray high-speed imaging was used to study the aggregate behaviour of ultrasonic cavitation bubbles, i.e., bubble cloud in a liquid alloy Sn-30%wtCu. A new method based on the X-ray attenuation for a white X-ray beam was developed to extract the quantitative information about the bubble clouds in the chaotic and quasi-static cavitation regions. This method is generic and applicable to all liquid metals. Statistical analyses were made on the bubble size distribution, and velocity distribution. Such rich statistical data provide more quantitative information about the characteristics of ultrasonic bubble clouds and cavitation in opaque, high-temperature liquid metals.

Author Contributions: C.W., T.C. and J.M. conducted the in-situ synchrotron X-ray experiment and led the writing of the paper. J.M., T.C. and D.E. were the Principal Investigators for securing the research funding from the UK Engineering and Physical Science Research Council. I.T. contributed to the writing and editing of the paper. All authors have read and agreed to the published version of the manuscript.

Funding: This research was funded by the EPSRC UltraCast project (Grant Numbers EP/L019884/1; EP/L019825/1; EP/L019965/1), and the award of synchrotron beamtime by Diamond Light Source, Beamline I12 (Experiment NT12131-1) are gratefully acknowledged.

Conflicts of Interest: The authors declare no conflict of interest. The funders had no role in the design of the study; in the collection, analyses, or interpretation of data; in the writing of the manuscript, or in the decision to publish the results.

References

1. Harvey, G.; Gachagan, A.; Mutasa, T. Review of High-Power Ultrasound-Industrial Applications and Measurement Methods. *IEEE Trans. Ultrason. Ferroelectr. Freq. Control* **2014**, *61*, 481–495. [CrossRef] [PubMed]
2. Mason, T.J. *Sonochemistry*; Oxfor University Press Inc.: New York, NY, USA, 1999.
3. Eskin, G.I.; Eskin, D.G. *Ultrasonic Treatment of Light Alloy Melts*, 2nd ed.; CRC Press: Boca Raton, FL, USA, 2014.
4. Eskin, D.G.; Mi, J. *Solidification Processing of Metal Alloys under External Fields*; Springer Nature Switzerland AG: Basel, Switzerland, 2018. [CrossRef]
5. Rayleigh, L. On the pressure developed in a liquid during the collapse of a spherical cavity. *Philos. Mag. Ser.* **1917**, *6*, 94–98. [CrossRef]
6. Ackerman, E. Pressure Thresholds for biologically active cavitation. *J. Appl. Phys.* **1953**, *24*, 1371–1373. [CrossRef]
7. Flynn, H.G. *Physical Acoustics*; Mason, W.P., Ed.; Academic Press: New York, NY, USA, 1964; Volume 1, pp. 58–172.
8. Lauterborn, W.; Kurz, T. Physics of bubble oscillations. *Rep. Prog. Phys.* **2010**, *73*, 106501. [CrossRef]
9. Zeqiri, B.; Gelat, P.N.; Hodnett, M.; Lee, N.D. A novel sensor for monitoring acoustic cavitation. Part I: Concept, theory, and prototype development. *IEEE Trans. Ultrason. Ferroelectr. Freq. Control* **2003**, *50*, 1342–1350. [CrossRef]
10. Harris, G.R.; Preston, R.C.; DeReggi, A.S. The impact of piezoelectric PVDF on medical ultrasound exposure measurements, standards, and regulations. *IEEE Trans. Ultrason. Ferroelectr. Freq. Control* **2000**, *47*, 1321–1335. [CrossRef]
11. Moussatov, A.; Granger, C.; Dubus, B. Cone-like bubble formation in ultrasonic cavitation field. *Ultrason. Sonochem.* **2003**, *10*, 191–195. [CrossRef]
12. Sijl, J.; Vos, H.J.; Rozendal, T.; de Jong, N.; Lohse, D.; Versluis, M. Combined optical and acoustical detection of single microbubble dynamics. *J. Acoust. Soc. Am.* **2011**, *130*, 3271–3281. [CrossRef]
13. Reibold, R.; Molkenstruck, W. Light-Diffraction Tomography Applied to the Investigation of Ultrasonic Fields. 1. Continous Waves. *Acustica* **1984**, *56*, 180–192.
14. Reibold, R. Light-Diffraction Tomography Applied to the Investigation of Ultrasonic Fields. 2. Standing Waves. *Acustica* **1987**, *63*, 283–289.
15. Price, G.J. *Current Trends in Sonochemistry*; The Royal Society of Chemistry: Cambridge, UK, 1992.
16. Price, G.J. ICA 2010—Incorporating Proceedings of the 2010 Annual Conference of the Australian Acoustical Society. In Proceedings of the 20th International Congress on Acoustics 2010, Sydney, Australia, 23–27 August 2010.
17. Ohl, S.W.; Klaseboer, E.; Khoo, B.C. Bubbles with shock waves and ultrasound: A review. *Interface Focus* **2015**, *5*, 15. [CrossRef] [PubMed]
18. Eskin, G.I. Principles of ultrasonic treatment: Application for light alloys melts. *Adv. Perform. Mater.* **1997**, *4*, 223–232. [CrossRef]
19. Wang, F.; Eskin, D.; Connolley, T.; Mi, J.W. Effect of ultrasonic melt treatment on the refinement of primary Al3Ti intermetallic in an Al-0.4Ti alloy. *J. Cryst. Growth* **2016**, *435*, 24–30. [CrossRef]
20. Mi, J.; Tan, D.; Lee, T. In Situ Synchrotron X-ray Study of Ultrasound Cavitation and Its Effect on Solidification Microstructures. *Metall. Mater. Trans. B* **2014**, *46*, 1615–1619. [CrossRef]
21. Tan, D.; Mi, J. *Light Metals Technology 2013*; Stone, I., McKay, B., Fan, Z.Y., Eds.; Trans Tech Publications Ltd.: Zürich, Switzerland, 2013; pp. 230–234.
22. Tan, D.; Lee, T.L.; Khong, J.C.; Connolley, T.; Fezzaa, K.; Mi, J. High Speed Synchrotron X-ray Imaging Studies of the Ultrasound Shockwave and Enhanced Flow during Metal Solidification Processes. *Metall. Mater. Trans. A* **2015**, *46*, 2851–2861. [CrossRef]

23. Wang, F.; Eskin, D.; Mi, J.; Wang, C.; Koe, B.; King, A.; Reinhard, C.; Connolley, T. A synchrotron X-radiography study of the fragmentation and refinement of primary intermetallic particles in an Al-35Cu alloy induced by ultrasonic melt processing. *Acta Mater.* **2017**, *141*, 142–153. [CrossRef]
24. Wang, B.; Tan, D.; Lee, T.L.; Jia, C.K.; Wang, F.; Eskin, D.; Connolley, T.; Fezzaa, K.; Mi, J. Ultrafast synchrotron X-ray imaging studies of microstructure fragmentation in solidification under ultrasound. *Acta Mater.* **2017**, *144*, 505–515. [CrossRef]
25. Furtauer, S.; Li, D.; Cupid, D.; Flandorfer, H. The Cu-Sn phase diagram, Part I: New experimental results. *Intermetallics* **2013**, *34*, 142–147. [CrossRef]
26. Tan, D. In Situ Ultrafast Synchrotron X-ray Imaging Studies of the Dynamics of Ultrasonic Bubbles in Liquids. Ph.D. Thesis, University of Hull, Hull, UK, August 2015.
27. Drakopoulos, M.; Connolley, T.; Reinhard, C.; Atwood, R.; Magdysyuk, O.; Vo, N.; Hart, M.; Connor, L.; Humphreys, B.; Howell, G.; et al. I12: The Joint Engineering, Environment and Processing (JEEP) beamline at Diamond Light Source. *J. Synchrotron Radiat.* **2015**, *22*, 828–838. [CrossRef]
28. Schneider, C.A.; Rasband, W.S.; Eliceiri, K.W. NIH Image to ImageJ: 25 years of image analysis. *Nat. Meth.* **2012**, *9*, 671–675. [CrossRef]
29. Gullikson, E.M. *X-ray Data Booklet*; Thompson, A.C., Ed.; Centre for X-ray Optics & Advanced Light Source: Berkeley, CA, USA, 2009; pp. 1–43.
30. Epanechnikov, V.A. Non-Parametric Estimation of a Multivariate Probability Density. *Theory Probab. Its Appl.* **1969**, *14*, 153–158. [CrossRef]

© 2019 by the authors. Licensee MDPI, Basel, Switzerland. This article is an open access article distributed under the terms and conditions of the Creative Commons Attribution (CC BY) license (http://creativecommons.org/licenses/by/4.0/).

MDPI
St. Alban-Anlage 66
4052 Basel
Switzerland
Tel. +41 61 683 77 34
Fax +41 61 302 89 18
www.mdpi.com

Materials Editorial Office
E-mail: materials@mdpi.com
www.mdpi.com/journal/materials

www.ingramcontent.com/pod-product-compliance
Lightning Source LLC
LaVergne TN
LVHW071959080526
838202LV00064B/6794